Mastering Instrument Flying

TAB
PRACTICAL
FLYING SERIES

Mastering Instrument Flying

Henry Sollman with
Sherwood Harris

TAB TAB BOOKS
Blue Ridge Summit, PA

Portions of this book's text and illustrations have been reproduced with permission of Jeppesen Sanderson, Inc.

No copyright claim is being made for any material which has been taken from U.S. Government sources.

The aeronautical charts and related data appearing in this book are for illustration purposes only and are *not* for use in navigation.

FIRST EDITION
FOURTH PRINTING

© 1989 by **Sherwood Harris and Henry Sollman**.
Published by TAB Books.
TAB Books is a division of McGraw-Hill, Inc.

Library of Congress Cataloging-in-Publication Data

Sollman, Henry.
    Mastering instrument flying / by Henry Sollman with Sherwood
  Harris.
        p.   cm.
    Includes index.
      ISBN 0-8306-9433-1   ISBN 0-8306-0433-2 (pbk.)
      1. Instrument flying.   I. Harris, Sherwood.   II. Title.
    TL711.B6S58   1989
    629.132′5214—dc20   89-32256
                                                           CIP

TAB Books offers software for sale. For information and a catalog, please contact
TAB Software Department, Blue Ridge Summit, PA 17294-0850.

Acquisitions Editor: Jeff Worsinger
Technical Editor: Norval G. Kennedy
Director of Production: Katherine G. Brown
Series Design: Jaclyn J. Boone
Cover photograph by Ray Duee.

Contents

Go or No-Go? Estimated Climb Out Time • Wind and Ground Speed • Flight Plan to Alternate • In-flight Notations • Filing the Flight Plan • Abbreviated Briefings • Outlook Briefings • The One-Call Technique • Total Time En Route • The Fast File Procedure

6 Airplane, Instrument, and Equipment Checks 65

Fuel Quantity • ATIS • Altimeter Error • Electrical Equipment • VOR Checks • Cockpit Organization • Radio Preparation • ILS Check • ADF Check • Transponder Check • Tips to Reduce Cockpit Confusion • Gyro Instruments

7 Clearances and Communications 73

Practice Clearances • Clearance Shorthand • Handling Amended Clearances • Obtaining Clearances • Unacceptable Clearances • Remote Communications Outlets (RCOs) • Void Time Clearances • Taxi Checks • Rolling Engine Run-up • Runway Checks • IFR Communications • Standard Phraseology • Who, Who, What, Where • Calling Ground Control • When You Hear Nothing Further • Managing Frequencies • Required Reports • Radio Contact Lost

8 Basic Instruments 85

First Instrument Flight • "Two, Two, and Twenty" • Overcontrolling • Attitude Control • Altitude Control • Control, Primary, and Support Instruments • Heading Control • SCAN Distractions • Support Instruments • Airspeed Control • Airspeed Transitions • Power • Instruments That Lie—When and Why

9 Turns, Climbs, and Descents 101

Turn Coordinator and Clock • Trim in a Turn • Oboe Pattern • Pattern A • Minimum Controllable Airspeed • Constant Airspeed Climbs • Departure Climb Out • Constant Rate Climbs • Chasing the Needle • Climb Level Off • Descents • Constant Speed Descents • Constant Rate Descents • Descent Level Off • Approach Descents • High Speed Final • Vertical S • Pattern B • Pattern C

10 VOR Procedures 117

Heading Indicator Errors • VOR Proficiency • 16-Point Orientation • VOR Time/Distance Check • Intercepting a Bearing or Radial • Wind Corrections • Common Interception Mistakes • Cleared Direct • Reference Heading • Bracketing • Chasing the Needle • Station Passage • Practice Patterns

11 Holding Patterns 131

Five Ts: Time, Turn, Twist, Throttle, Talk • Wind Corrections • En Route Holding • Holding Pattern Entry • Choosing the Correct Entry • Importance of Altitude Control • Holding Pattern Variations • Intersection Holds • One-VOR Intersection Holds • DME Holding Patterns

Acknowledgments

My collaborator on *Mastering Instrument Flying* was Sherwood Harris, a veteran pilot who was a senior editor in the books divisions of **Reader's Digest** for many years. His unique combination of talents as a flight instructor, author, and editor made this book possible. He found the words that made my ideas come to life on paper.

I would also like to thank all my colleagues at Panorama Flight Service Inc. for their support and assistance and especially my assistant, John Jenkins, for his proofreading and suggestions. The instructors used my syllabus for training, and provided many important insights with their comments and suggestions.

Finally, thanks to all my students down through the years. A good teacher always learns from his students and mine have taught me very well indeed.

—Henry Sollman

How to Use This Book

As a flight instructor in the 1960s, I kept running across more and more students who had tried without success for months—sometimes years—to get an instrument rating. They would fly countless hours with their instructors, but they never seemed to get anywhere. They couldn't see the light at the end of the tunnel!

So I started developing the concepts for an accelerated instrument course, a 10-day program that would train my students to be safe and confident instrument pilots and give them everything they needed to pass their instrument flight tests with ease. I believe I was the first flight instructor to offer "an instrument rating in 10 days," a course that really worked. The course requires "total commitment" on the part of the student and instructor. And, if the student intends to complete the course within 10 days, he must show up with a passing grade on the instrument written test.

I placed three ads for my 10-day instrument course in *Trade-a-Plane*. The response was so great that I had to stop running the ads. I even had people from Europe applying for the course, and, in due course, I enrolled my first foreign students: two pilots from Hamburg, West Germany.

Since that time thousands of pilots—American and foreign—have benefited from my system of teaching instrument flying. I have had many imitators, but I believe that no one has duplicated the quality and performance that I have achieved in more than 20 years training instrument pilots.

I present the basic building blocks of instrument flying as the student actually encounters them in the real world of IFR. We work on flight planning, weather, and clearances first, then proceed to the fundamentals of controlling the airplane by instruments, then to VOR and ADF procedures, then to approaches, and so forth. When an instrument student works with me, we never go up and just bore holes aimlessly in the sky!

On all instrument training flights, I always have my students prepare and file a complete IFR flight plan to a real destination with return, even though we might cancel IFR en route to practice basic instrument flying, holding patterns or instrument approaches to accomplish the objectives of the lesson.

My insistence on planning and filing IFR on every training flight is beneficial. Students quickly become adept at getting good weather briefings, making real go/no-go weather decisions, and choosing appropriate alternate airports. In a surprisingly short time, they become expert at copying clearances and working like professionals with air traffic control (ATC) on instrument departures, en route procedures, and instrument approaches.

Working with ATC on instrument flights is a matter of repetition and routine, and is best learned through practice. The sooner the ATC procedures are introduced, the sooner the student becomes competent and confident. And the sooner the student becomes competent and confident in ATC procedures, the sooner he can concentrate his attention on other important aspects of instrument flight. A student who stumbles over clearances at the beginning of an instrument flight is just that much more likely to stumble over his dealings with ATC on an instrument approach.

Last, and perhaps not least for many people, going through these routines on each and every training flight gives my students a much better chance for success on the instrument flight test. I am also a designated examiner, and I can tell you that an applicant who has any difficulty with weather, flight plans, or copying clearances has little or no chance of obtaining an instrument rating from me, or anyone else. Who wants to share airspace on an IFR day with someone who has not thoroughly mastered these important fundamentals?

Over the more than 50 years that I have been flying and instructing, I have learned many tricks of the trade that help me teach and help my students learn. Many of these are what I call "shortcuts that don't short-change." These special techniques—honed and polished over many years with all kinds of students—are important ingredients in the success of my teaching. All are included in this book.

Finally, I have never allowed my course to stagnate. Aviation is a dynamic, constantly changing field, and recently no area of aviation has changed more rapidly for the general aviation pilot than air traffic control and IFR procedures.

This book is an example of my constantly evolving approach to flight instruction. *Mastering Instrument Flying* is an entirely new course designed from beginning to end to meet or exceed the new instrument flight test standards published by the Federal Aviation Administration in 1986. The course has been tested and "debugged" over many months of actual use by the instructors on my staff at Panorama Flight Service.

Mastering Instrument Flying is:

1. A basic course for instrument students. It contains the first syllabus of 20 flight lessons designed specifically to fulfill the requirements of the redesigned FAA practical test standards.

Supporting the syllabus are chapters that cover, in detail, the material that should be read and understood prior to each flight lesson. This material has been carefully illustrated to make everything absolutely clear. Flight lessons are cross-referenced for the required reading for that flight lesson.

2. A set of lesson plans, with explanations, for the instrument flight instructor. The syllabus is designed for completion within two weeks if both instructor and student can devote full time to the course. If this is not possible, there is sufficient reinforcement and review built into the 20 lessons to qualify a student for an instrument rating, with no wasted time or effort in a less intensive pace.

3. A handy refresher for the pilot with an instrument rating who needs to brush up on procedures while preparing for an instrument competency check.

Throughout the book the emphasis is on practical problems of instrument flight, rather than theoretical problems. Reference material regarding aerodynamics, weather theory, instrument interpretation, radio signal propagation, and the like abound. Instrument students, pilots, and instructors should be thoroughly familiar with these sources. The best are compiled in Appendix A, The Instrument Pilot's Professional Library. Also provided is information on how to obtain these publications—a process that is not always as easy as it should be.

Appendix B contains excerpts from Federal Aviation Regulations (FARs) 1, 61, and 91 that deal with instrument training and IFR flight.

The Glossary contains pilot and controller terms encountered when planning, filing, and flying an IFR flight.

1
The Psychology of Instrument Flight

Mastering instrument flying is as much a state of mind as it is training, study, and practice.

FROM THIS MOMENT FORWARD IT IS IMPORTANT THAT YOU ESTABLISH THE GOAL OF being the "Proud, Perfect Pilot!" Each time you fly, aim to be as perfect as possible in everything you do. Be a nitpicker and be as tough with yourself as you possibly can.

Don't try to master everything all at once. Instead, try to be perfect with just one element and build on the previous skills you have learned. Fly relaxed but push yourself to the very limit. If you find yourself tiring or feeling overburdened, stop and relax.

How do you become a proud, perfect pilot? Let's start with the preflight preparation. Determine what you want to accomplish in your preflight planning and do it over and over again until you can do it perfectly. Concentrate on one thing at a time—the one thing that you want to do perfectly at that moment.

Try to master the planning involved in filling out the flight log. Do everything you possibly can to understand the flight you expect to fly. Be a nitpicker with the details until you can plan a flight perfectly to your satisfaction. Then get an evaluation from your flight instructor and incorporate the instructor's recommendations. Then do it again and again until it is perfect. (In the legal profession, preparation is 90 percent of the law. The same can be said for a perfect instrument flight.)

Plan alternate courses of action in case ATC might route you differently from what you planned. It is fun to try to outwit ATC and understand what goes on in their minds and why. There must be a reason! Work out a detailed flight log for alternate courses; it will be time well spent.

THE PSYCHOLOGY OF INSTRUMENT FLIGHT

Considerable attention in your early training should be given to the basics. You can't be a perfect instrument pilot unless you can control heading and altitude. Concentrate on heading control until you can maintain a heading within ±2 degrees; don't worry about altitude.

Concentrate on altitude until you can maintain altitude within ±20 feet; don't worry about heading. Individually, you will master these basics rather quickly. Then all you have to do is put them together. When you can do this you will have mastered a skill that you will use throughout future flights, whether VFR or IFR. Just think about how much fun it is to be perfect!

2
Mastery of Instrument Flight in 20 Lessons

MASTERY OF THE 20 STEP-BY-STEP LESSONS AND THE FOUR BACKGROUND BRIEFINGS in the Instrument Rating Syllabus (Chapter 18) allows an instrument student to become proficient in all procedures and maneuvers needed to become a safe and confident instrument pilot. The syllabus also ensures coverage of all the tasks required by the new *Instrument Rating Practical Test Standards* to pass the FAA instrument pilot (airplane) flight test.

FIGURE 2-1 is reprinted from the new *Instrument Rating Practical Flight Test Standards*. It is a checklist of the areas of knowledge and skill, or tasks, in which an applicant must demonstrate acceptable performance before an instrument rating will be issued. (Yes, the examiner really must evaluate all tasks on an instrument flight check!)

Next to each task is a checkbox that can be used to record a student's progress through the syllabus. I suggest that the student make one slash from corner to corner through the box when he or she believes that a task has been mastered. The instructor can make another slash going the other way when satisfied that the student's performance meets FAA standards for passing the flight test, and then write in the date. Not only will this provide a record of the student's progress, but it will also encourage students to analyze their own performance and compare what they think of it with how their instructor evaluates them.

Take a moment now to note how the material in "The Instrument Rating Syllabus" (Chapter 18) is presented. Each flight lesson and background briefing is numbered in sequence. Each has an introduction.

There is assigned reading from this book that must be completed prior to each flight lesson. The assigned reading provides the information you must have to understand and

Fig. 2-1. *Checklist of tasks that are required for the issuance of an instrument rating, as published in the FAA's* Instrument Rating Practical Test Standards.

I. GROUND PHASE

- ☐ **A.** Obtaining weather information.
- ☐ **B.** Cross-country flight planning.
- ☐ **C.** Aircraft systems related to IFR operations.
- ☐ **D.** Aircraft flight-instruments and navigation equipment.
- ☐ **E.** Instrument cockpit check.

II. AIR TRAFFIC CONTROL CLEARANCES AND PROCEDURES

- ☐ **A.** Air traffic control clearances.
- ☐ **B.** Compliance with departure, en route, and arrival procedures, and clearances.
- ☐ **C.** Holding procedures.

III. FLIGHT BY REFERENCE TO INSTRUMENTS

- ☐ **A.** Straight-and-level flight.
- ☐ **B.** Change of airspeed.
- ☐ **C.** Constant airspeed climbs and descents.
- ☐ **D.** Rate climbs and descents.
- ☐ **E.** Timed turns to magnetic compass headings.
- ☐ **F.** Steep turns.
- ☐ **G.** Recovery from unusual flight attitudes.

IV. NAVIGATION AIDS

- ☐ **A.** Intercepting and tracking VOR/VORTAC radials and DME arcs.
- ☐ **B.** Intercepting and tracking NDB bearings.

V. NO-GYRO RADAR PROCEDURES

- ☐ **A.** No-gyro radar vectoring and approach procedure.

VI. INSTRUMENT APPROACH PROCEDURES

- ☐ **A.** VOR/VORTAC instrument approach procedure.
- ☐ **B.** NDB instrument approach procedure.
- ☐ **C.** ILS/MLS instrument approach procedure.
- ☐ **D.** Missed approach procedures.
- ☐ **E.** Circling approach procedure.
- ☐ **F.** Landing from a straight-in or circling approach procedure.

VII. INSTRUMENT FLIGHT IN A MULTIENGINE AIRCRAFT

- ☐ **A.** Engine failure during straight-and-level flight and turns.
- ☐ **B.** Instrument approach—all engines operating.
- ☐ **C.** Instrument approach—one engine inoperative.

perform the pilot tasks prescribed for the flight lesson. If you are not familiar with the assigned reading, it will take you much longer—and cost you a lot more—to complete the flight lesson.

Note also that each flight lesson is preceded by a preflight briefing to cover the objectives of the flight lesson. Likewise, each flight lesson is followed by a postflight critique to assess the flight and to discuss the next lesson. Flight lessons contain both review and new material, and completion standards are also given so that the student's progress can be measured in terms of the levels set in the *Instrument Pilot Practical Test Standards*.

There is no set time allotted for each of the 20 lessons in the syllabus that follows. A lesson is complete only when the completion standards are met, not when a specified amount of time has been devoted to it. Instructors may repeat lessons or portions of lessons as needed for the student to meet the completion standards.

SUPPLEMENTARY EXERCISES

A special feature of the syllabus is maneuvers designated Supplementary Exercises in many of the flight lessons. Supplementary exercises provide extra practice to help students cure common problems encountered in instrument training, or to improve their proficiency with advanced maneuvers.

Mastery of the supplementary exercises is not necessary to meet the completion standards for a flight lesson, and they are not required on the flight test.

Student Note. If you would like to get extra practice on any of the required instrument maneuvers or the supplementary exercises without an instructor on board, you should take advantage of every opportunity to do so. The more you practice the maneuvers, the more competent you will become. You must fly with a safety pilot.

SAFETY PILOTS

Regulations require a second safety pilot to watch out for other traffic, clouds, and obstacles, such as radio towers and hills. According to FAR 91.21(b), "no person may operate a civil aircraft in simulated instrument flight unless an appropriately rated pilot occupies the other control seat as a safety pilot."

"Appropriately rated" means holding a pilot certificate with category and class ratings appropriate for the aircraft used. If you are acting as safety pilot in a Cessna 172, for example, your pilot certificate must have an airplane single-engine land rating; if you are in an Aztec, your pilot certificate must have an airplane multi-engine land rating. A safety pilot must also have a current medical certificate because the safety pilot is a "required pilot flight-crewmember" in terms of FAR 61.3(c).

FLIGHT EXPERIENCE REQUIREMENTS

As students proceed through the syllabus with an instructor, they will work toward the FAA "flight experience" requirements for an applicant for an instrument rating as stated in FAR 61.65, which are:

(1) A total of 125 hours of pilot time, of which 50 hours are as pilot in command in cross-country flight in a powered aircraft with other than a student pilot certificate. Each cross-country flight must have a landing at a point more than 50 nautical miles from the original point of departure.
(2) 40 hours of simulated or actual instrument time, of which not more than 20 hours may be instrument instruction by an authorized instructor in an instrument ground trainer acceptable to the Administrator.
(3) 15 hours of instrument instruction by an authorized flight instructor, including at least five hours in an airplane or a helicopter, as appropriate.

In addition, for the airplane instrument rating, FAR 61.65(c)(4) requires "cross-country flying in simulated or actual IFR conditions, on federal airways or as routed by ATC, including one such trip of at least 250 nautical miles, including VOR, ADF, and ILS approaches at different airports."

Simulated instrument time should be logged by the student only when operating solely by reference to instruments while using a hood or other device that blocks outside visual references. Actual instrument time can be logged only when operating solely by reference to instruments or under instrument meteorological conditions (IMC) when neither the horizon nor the ground is visible because of the weather. The student and the instructor may both log actual instrument time simultaneously, but only the student may log simulated instrument time because the instructor will be "visual" while acting as safety pilot.

LOGGING PILOT-IN-COMMAND TIME

The student may log pilot-in-command time while sole manipulator of the controls in simulated or actual IMC. The instructor may log pilot-in-command time for the entire training flight.

Instructor Note. You should conduct as much flight training as possible in actual IFR conditions. Training flights in actual IFR weather will prepare a student for the real world of instrument flying encountered after receiving an instrument rating. If the weather is too adverse for a satisfactory training flight, carry out the lesson in an FAA-approved instrument ground trainer.

When the weather is VFR students must use a hood or other approved view limiting device and log flight time as simulated instrument time. Remember, the examiner will expect the candidate for an instrument rating to use an approved view limiting device throughout the entire check flight.

PREFLIGHT AND POSTFLIGHT BRIEFINGS

To get the greatest benefit from the syllabus flight lessons, each lesson must include a preflight briefing and a postflight critique.

The preflight briefing will cover the objectives of the lesson, the procedures and maneuvers to be practiced, and standards the student is expected to achieve.

The postflight critique will clearly establish the elements of the lesson that the student has mastered, as well as those that need further attention. The postflight critique will also include a preview of the next flight lesson and, because every flight lesson will start by filing an IFR flight plan, the instructor will suggest the destination for planning and filing IFR on the next lesson.

Instructor Note. All training IFR flight plans must be filed in the instructor's name according to FAR 61.3 (e)(1).

Students will progress much more rapidly (and at considerably less cost!) if they are thoroughly familiar with the "Assigned Reading" prior to each flight lesson and background briefing. The reading assignments are referenced here by chapter number, chapter title, and page.

Assigned reading describes in detail what will take place during each flight lesson. The reading material also contains information and background that must be understood in order to successfully complete the flight lessons.

BACKGROUND BRIEFINGS

Students should note the inclusion of four background briefings prior to four chapters that introduce especially important new material: basic instruments, approaches, IFR cross-country flights, and the instrument flight test. Students should prepare for background briefings by studying the assigned reading on their own and writing answers to all the questions. A few review questions also appear in each background briefing. These are review questions are based on material previously assigned as required reading for the flight lessons, or based on material covered in VFR training that must be re-emphasized in IFR training.

There are no trick questions in the background briefings, however, it may take some research to find satisfactory answers. That is exactly what the background briefings are meant to do—make you dig for the information. What you learn by researching will stick with you much longer than material learned by rote.

That's also the reason why no answers are supplied in this book; if the answers were readily available, there would be no incentive to dig out the information on your own. Your instructor will have the answers and will provide them as you discuss the background briefings.

As soon as you answer all the questions in one background briefing, start working immediately on questions in the next briefing. You want to be prepared to go over the briefings with your instructor at the appropriate points in the syllabus, and it does take time to prepare all the answers!

Instructor Note. Be prepared also! Be sure you have all the answers under control before you go over the briefings with your students so you can discuss all the questions and answers intelligently. All the reference material is not in this book. Many of the references cited are other readily available publications listed in Appendix A, "The Instrument Pilot's Professional Library."

You must do your homework on the background briefings before your student's lesson. Answers won't be given verbatim in the reference material! You may have to work out satisfactory answers completely on your own for some of the questions.

You should encourage students to obtain the necessary publications in Appendix A on their own. But if a student has not yet had time to obtain them, you should be prepared to lend the publications to the student from either your own professional library or from that of the flight school. You might want to prepare handouts for students by photocopying the relevant pages.

Do not confuse background briefings with preflight briefings and postflight critiques. These are conducted at the beginning and end of every flight lesson. Background briefings are separate sessions between flight lessons. They contain too much material to be included at the beginning or end of a flight lesson.

Background briefings are numbered to show when they should be conducted. For example, Backgroung Briefing 1-2 comes after Flight Lesson 1 but before Flight Lesson 2.

The student and instructor should schedule ground school time to cover the material in the background briefing prior to the flight lesson containing new material. The ground school sessions should last two or three hours and may be scheduled over two or three days.

Students and instructors should go through the briefings together. Instructors should make certain that students clearly understand the material covered by the assigned reading and all the questions before proceeding with the next flight lesson.

3
Preparing for an Instrument Flight

PLANNING AN IFR FLIGHT SEEMS TO BE A LARGE ORDER FOR THE BEGINNING INSTRUMENT student. There are new charts to master, the weather analysis is more complex compared to VFR analysis and procedures for the departure, en route, and approach phases of the flight may be unclear.

It's only natural to be a little puzzled at first. However, there are many things that can be done to master the process more quickly and make it more interesting; yes, believe it or not, flight planning can even be fun!

The single most important factor in taking the mystery out of IFR flight planning is to plan and file an IFR flight plan on every training flight, including the very first one.

Even if you are only going out to practice Pattern A in the local area, you should still plan and file IFR to a destination 50 to 75 nautical miles from the home airport. You should also work out an IFR flight plan for the return trip, whether you expect to use it or not.

In the beginning the instructor will suggest two or three nearby destinations. Get the instructor's ideas well ahead of the flight so you can plan at home when you have more time and you are under less stress.

Plan for all destinations the instructor has suggested, not just one. This will give you more practice, it will expose you to a variety of situations, and it won't cost you a cent! Students who show up at the airport without their homework finished waste both their time and their instructor's time.

You will be surprised how much you can accomplish at home. You can map out the flight route, review departure procedures, and go over the destination approach charts and airport information in great detail. You can call from home, get preliminary weather

Fig. 3-1. *Flight log form for planning and logging IFR cross-country flights.*

FIX		INTERSEC	ROUTE	M.C.	DIST.	G/S	TIME ENR	TIME ARR	DATE
	NAME	NAME	VIA	TO	TO	EST	EST	EST	FLIGHT PLAN ROUTE
		IDENT							
		FREQ							
IDENT.	FREQ	RADIAL	ALT	FROM	REM	ACT	ACT	ACT	
									ATC CLEARS N _____
									CLEARANCE LIMIT
AC ID	POSITION	TIME	ALT.	IFR/VFR	EST. NEXT FIX		NEXT FIX		PIREPS

PILOT:

			CALL SIGN	FREQ.	ASGND ALT
TAS:	IAS:				
RPM:	MP:				
TAKEOFF: RWY | WIND | KTS | ALT			CPT		
			GRND		
			TWR		
Field Elev:			DC		
ALTIMETER ERROR		[ ]			
LANDING:					
Field Elev: RWY | WIND | KTS | ALT					
TYPE OF APPROACH:					
TIME ON:		[ ]			
TIME OFF:					
TOTAL TIME ENROUTE:			AC		
TACH ON	TACH OFF		TWR		
			GRND		

WINDS ALOFT

Sta.	A	Dir.	Vel	Temp	A	Dr	Vl	T	A	D	V	T

VFR WX AT:

TANK ON	TANK OFF		TANK ON	TANK OFF

briefings, check the NOTAMS, and even file your flight plan. In fact, you can make several calls if the weather is changing rapidly, or there is something you didn't catch clearly on the first call. This is particularly helpful in the beginning of your IFR training.

ATC will often make amendments to your clearance as the IFR flight proceeds. When you carefully work out your flight at home, familiarize yourself with other VORs and airways between your departure and destination airports, and pick out all those obscure intersections that ATC may use for clearance limits, re-routing, or holding fixes.

The cockpit is not the place for basic research. A newly-rated instrument pilot departed one of the New York airports in actual IFR weather flying a light twin with his family on board. He had only filed and flown IFR a few times before he received his rating. When airborne, he contacted Kennedy departure control. Kennedy responded: "New clearance. Ready to copy?" It wasn't a major change, but due to his inexperience, it overloaded him. He lost control of the airplane and it crashed, killing all on board. If ever an accident could have been prevented by more thorough training, this was it.

THE FLIGHT LOG

As a VFR pilot you already have considerable experience planning cross-country flights and navigating with VORs. Your IFR planning is nothing more than an extension of what you have already learned, with more detail in some areas and greater emphasis on other points that are not so crucial to VFR flying. Over the years I developed a flight log form (FIG. 3-1) that covers all information needed for both VFR and IFR cross-country flights. I find that if my private pilot students use this log form for the cross-country phase of their training, they make the transition from VFR to IFR flight planning very easily.

So be my guest—copy the form and use it for your VFR and IFR flying. It gives you a logical, step-by-step way to work your way through all the important elements of flight planning. If you fill out the log properly, nothing has been overlooked.

During the flight itself, the form functions as a running log with places to enter time en route, ETA, actual time of arrival, ground speed between fixes, and fuel consumption. Equally important, it provides a quick reference for all the detailed information you need to conduct the flight without a lot of confusion and fumbling around. Cockpit organization is the key to a good IFR flight.

Let's go through the form step-by-step so you can see how useful it is when planning a flight. We'll be taking a hypothetical IFR cross-country from Westchester County Airport, White Plains N.Y. to Broome County Airport, Binghamton N.Y., on a typical actual IFR day in late November. I have chosen this example because I have found that this trip to Binghamton contains all the elements of a good training flight. The principles and techniques that apply here are valid everywhere else.

IFR flight planning occurs in two phases. In the first phase you decide your route of flight and fill in your flight log with all the information available before obtaining a weather briefing. In the second phase, get your weather briefing, then fill in the information affected by the weather, such as ground speed and the choice of an alternate airport.

PLANNING THE ROUTE

The first step is plotting the route between your departure and your destination airports, including the appropriate Standard Instrument Departure (SID), if there is one. Open up the en route low altitude chart, or charts, that cover the area—in this case Chart L-25. Notice what Victor airways make a logical route to your destination; V252 from Huguenot VOR goes directly to Binghamton VOR, so that is the best Victor airway for the en route portion of this flight (FIG. 3-2).

You could also get from Huguenot to Binghamton by departing Huguenot on V126 to Lake Henry, then taking V149 to Binghamton. This would lengthen the trip unnecessarily so V252 is the better choice. However, on many IFR cross-country flights the best route to your destination may involve two or more Victor airways, and there is nothing wrong with combining several. Also, an airway segment may have several Victor airway numbers. The airway heading northeast from Sparta on L-25 is numbered V205-249 above the line and V89 below. Which one would you list on your log and flight plan? The answer: either one. It doesn't matter as far as ATC is concerned. Pick the numbered route that will carry you farthest toward your destination.

STANDARD INSTRUMENT DEPARTURES (SIDs)

Note that there is no Victor airway from Westchester Airport to Huguenot. How do you get to Huguenot? Utilize Westchester's published Standard Instrument Departure (SID) called "Westchester 9" (FIG. 3-3). If there is no SID available at your departure airport, you would simply file "direct" to the first VOR on the airway you intend to use.

Your IFR flight planning should always include a study of the SIDs available at your departure airport because ATC will expect you to use them. SIDs are published by the government in two booklets that cover the eastern and western regions of the United States. SIDs are also published by Jeppesen and are included as separate sheets along with the Jeppesen approach charts for an airport.

PREFERRED ROUTES AND TECs

Before you enter the planning results on the flight log, check two more items that often can be very helpful. They are Preferred IFR Routes and Tower En route Control (TEC). (In the beginning of a student's instrument training, I don't emphasize Preferred IFR Routes and TECs because they can be confusing.) You should know about them and become familiar with them.

Preferred IFR Routes are listed toward the back of the Airport/Facility Directory (AFD)—the green book. As the name implies, Preferred IFR Routes have been established by the FAA to guide pilots in their flight planning and to minimize route changes during the flight. If a preferred route doesn't cover your entire flight, use whatever segments you can. You are more likely to get the route you file for and less likely to have the route amended in the air if you can use all or part of a preferred route.

Fig. 3-2. *The route discussed in text from Westchester County Airport, New York, to Binghamton, New York, as plotted on NOS En route Low Altitude Chart L-25. Also shown is the route to the alternate, Wilkes-Barre, Pennsylvania.*

Fig. 3-3. *Westchester Nine Departure, the standard instrument departure (SID) from Westchester County, N.Y., Airport, as depicted in the NOS Standard Instrument Departures booklet covering the Eastern United States.*

TEC routes and airports are also listed toward the back of the AFD. TEC makes it possible to fly from one approach control point to another without entering Air Route Traffic Control Center (ARTCC) airspace. Designed primarily for instrument flights below 10,000 feet, TEC is usually quicker and simpler than routing through center airspace. But it's available only between the paired airports listed in the AFD; if your departure airport is not paired with your destination airport, you cannot use TEC. If you can use a TEC route, enter the acronym TEC in the remarks section of your flight plan.

Browse through both the preferred routes and the TEC sections of the AFD. Examine how the information is set up and check to see if your usual departure point fits into either or both categories. If so, these two special procedures will simplify your planning and save time.

NOS OR JEPPESEN CHARTS?

I'm often asked which instrument charts I prefer, those published by Jeppesen, for a loose-leaf notebook, or those published by the government, the National Ocean Service (NOS) booklets of instrument approach procedures charts, SIDs, and the en route charts.

Both are equally good and authoritative, and for the experienced instrument pilot, it all boils down to a matter of personal preference. However, for training purposes I prefer the government charts. For one thing all the examples used in the instrument written test are chosen from NOS and en route charts. I think it is less confusing to use only one type of chart throughout all phases of instrument training.

I believe that the NOS Instrument Approach Procedures chart booklets are a little easier for the instrument student to cope with. When the charts expire, you merely toss out the old booklet and substitute the current booklet. With a Jeppesen subscription you get a packet of pages to replace in your ring binder every two weeks. It's a temptation to put off the chore of inserting the pages until the day of the flight, when the time spent inserting the pages could be put to better use on other aspects of IFR flight planning.

APPROACH PLANNING

If you have worked out the route to your destination, you should study the Standard Terminal Arrival charts (STARs), if any apply, as well as the Instrument Approach Procedure charts for your destination airport. STARs help you plan your transition from the en route phase to the approach phase. Like SIDs, STARs are issued by NOS in separate booklets.

Look for two things when you study STARs and approach charts. First you need to familiarize yourself with *all* the approaches available at the destination because you must be prepared for the approach that ATC assigns. In the beginning of your instrument training, the amount of detail packed on these small charts may be quite mystifying. For example, there are no less than five different instrument approach procedures to Binghamton Airport!

No one expects you to be fully cognizant of all the approach procedures on your first few training flights. But you must examine all the approaches for your destination every time you file a flight plan. You will be amazed at how much information you will be able to absorb after you do this a few times.

Don't forget your departure airport; familiarize yourself with its instrument approaches in case a door pops open or a passenger becomes sick or for some other reason you have to return immediately after takeoff.

Airport Diagrams. Study the airport diagrams first, just as you would on a VFR cross-country. Note the field elevation. Study the runway diagram and note how many runways, their headings, lengths, and widths. Will you be able to use all runways? Do mountains or other obstructions affect an instrument approach? How high are the obstructions? Check over the taxiways because getting from your runway to your destination on the field can sometimes be the most complicated part of the trip. (One time at Montreal taxiing to the FBO was longer than the flight from Ottawa!)

Initial Approach Fixes (IAFs). Look at all the VORs, NDBs, marker beacons, intersections, and holding patterns. How will you get from the Victor airway you've been flying to the most likely instrument approach? Ask yourself which approach you will probably be assigned when you are handed off to approach control; consider the prevailing wind. Then pick out the VOR, NDB, or intersection on your inbound route that is closest to the final approach course. Chances are, this fix will be designated as an Initial Approach Fix (IAF) on your approach chart.

For our hypothetical flight to Binghamton, the wind is generally northwesterly, so the most likely approach is ILS RWY 34 (FIG. 3-4). We could simply continue on V252 to Binghamton VOR and use that as our destination fix. But that would involve a lot of maneuvering to get back to the final approach course for Runway 34.

A better choice for destination fix would be Latty intersection. Latty is on both V252 and the final approach course for Runway 34 and it appears on both the L-25 chart and the approach chart for ILS RWY 34. In most cases, approach control will give you radar vectors to the final approach course for the approach.

But if there is no radar available, or you have lost radio communications, you will be expected to make the transition from the Victor airway to the final approach course on your own. Beware—your instructor may insist that you make the full approach! So you must plan ahead.

Now we are ready to fill out the flight log. In the "Flight Plan Route" section at the center top portion of your flight log form, write in the abbreviations that will spell out your planned route, in this case—WESTCHESTER 9 SID, HAAYS INTER-SECTION, DIRECT HUGUENOT, V252 LATTY INTERSECTION (FIG. 3-5). This is also entered in Block 8 of your flight plan form for the "Route of Flight."

Amdt 21 88322

ILS RWY 34
AL-20 (FAA)

BINGHAMTON/EDWIN A. LINK FIELD-BROOME COUNTY (BGM)
BINGHAMTON, NEW YORK

BINGHAMTON APP CON
118.6 397.2
BINGHAMTON TOWER
119.3 242.7
GND CON
121.9
ASR
ATIS 128.15

LOCALIZER 110.3
I-BGM

BINGHAMTON
112.2 BGM
Chan 59

275°
095°
R-275

3900
116°
(11.4)

△ GROWS
3900 227°
(14.9)

1687±
1750±
MM

SMITE
OM

R-116

NIMMONS
INT

2564 ΛΛ

R-132

IAF
NIMMONS
332 IMZ

HANCOCK
116.8 HNK
Chan 115

3900
340° (7.3)

R-272

LATTY

4200 NoPT
345° (34.7)

10 NM

FEEDER FACILITIES

R-003

(IAF)
LAKE HENRY

MSA IMZ 25 NM ENROUTE FACILITIES

3800

WILKES BARRE
111.6 AVP
Chan 53

MISSED APPROACH
CLimb to 2200 then climbing
left turn to 3900 direct BGM
VORTAC and hold.

NDB/INT
3803

One Minute
Holding Pattern

OM
2846

MM

160° → 3900
←340°

3900

GS 3.00°
TCH 57

0.5 ├─3.3 NM──┤──3.1 NM──┤

ELEV 1630

9ι (Λs) Λ 1705
Λ 1638
1690
TWR
1678 ☆
4999 X 150
1692 Λ
1592
TDZE
1597

340° 6.9 NM
from FAF

CATEGORY	A	B	C	D
S-ILS 34	1797/24 200 (200- ½)			
S-LOC 34	1960/24 363 (400-½)			1960/40 363 (400-¾)
CIRCLING	2060-1 430 (500-1)	2080-1 450 (500-1)	2080-1½ 450 (500-1½)	2180-2 550 (600-2)

Cat. D S-LOC visibility increased to RVR 5000 for inoperative MM.

MIRL Rwy 10-28
HIRL Rwy 16-34

FAF to MAP 6.9 NM					
Knots	60	90	120	150	180
Min:Sec	6:54	4:36	3:37	2:46	2:18

ILS RWY 34

42°12'N-75°59'W

BINGHAMTON, NEW YORK
BINGHAMTON/EDWIN A. LINK FIELD-BROOME COUNTY (BGM)

Fig. 3-4. The ILS 34 approach to Binghamton as shown in the U.S. Northeast booklet of NOS Instrument Approach Procedures. Note the location of LATTY Intersection, the final fix on the IFR flight to Binghamton discussed in the text.

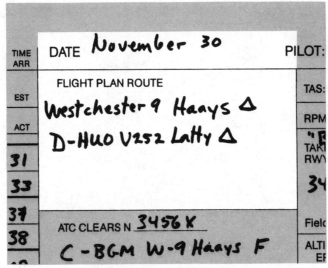

Fig. 3-5. *Flight log form with "Flight Plan Route" block filled in for flight to Binghamton.*

EN ROUTE FIXES

Now let's fill in more details of your flight. Look at the eight columns in left one-third of the flight log (FIG. 3-6). No mysteries here, identical to a VFR flight with a few extra details. Spell out the full name of all VORs used for fixes along your route. You may not be familiar with the three-letter identifiers, at least not in the beginning, and the names of some VORs can be very confusing.

For example, who could possibly guess that the identifier HPN stands for my homebase Westchester County Airport in White Plains, New York? So to avoid confusion, write out the station name as shown with its identifier and frequency in the boxes below.

If your fix is an intersection, use the column under "intersec," enter the name of the intersection as shown and list the identifier, the frequency, and the radial identifying the intersection. Note that we have entered LATTY intersection here because of its usefulness in locating the final approach course and it is common to both the en route chart and ILS 34 at Binghamton.

The next three columns are for routes, altitudes, magnetic courses, and distances; fill in as shown. The altitude in the third column is what you shall request for the first leg of the flight. The ground speed, time en route, and time of arrival (time arr) columns will have to wait until your weather briefing provides wind information at your planned altitude.

Actual ground speeds, actual times en route, and actual times of arrival will be entered as you proceed with the flight. This versatile form not only provides a structure for your flight planning but also becomes a very handy log to monitor the progress of your flight.

FIX		INTERSEC	ROUTE	M.C.	DIST.	G/S	TIME ENR
NAME		NAME	VIA	TO	TO	EST	EST
		IDENT					
IDENT.	FREQ	FREQ RADIAL	ALT	FROM	REM	ACT	ACT
	–	Haays STW	W-9	305	37+	108	26
HUO	116.1	109.6 054		325			28
Huguenot			V273	325	8	144	4
HUO	116.1			314			5
		Lany HNK	V252	318	66	134	30
BGM	112.2	116.8 272					28
BGM -airport			ILS	340	14	109	8
I-BGM	110.3						9
						125	68
							(1+08

Fig. 3-6. *Fixes, magnetic courses, and mileages entered for flight to Binghamton.*

COMMUNICATIONS FREQUENCIES

A few more details can be filled in before you get your weather briefing. Look up your departure and destination airports again in the A/FD and enter the frequencies as shown in the three columns at the far right of the form (FIG. 3-7). If your departure airport has an Automatic Terminal Information Service (ATIS), write in this information first at the top of the first column.

The abbreviations for the other entries are: CPT, Clearance Prior to Taxi; GRND, Ground Control; TWR, Tower; and DC, Departure Control. Use the blank boxes to list your en route frequencies as they are assigned to you.

The column on the far right is reserved for any changes of assigned altitudes. A typical change of en route frequencies might be: "Contact New York Center now, one two eight point five." When you make the frequency change and New York Center advises "descend and maintain four," you simply write in the new altitude next to the frequency. Enter NY CTR 128.5 40 (for 4,000 feet) in the appropriate boxes. Once again the flight log form functions as a flight planning guide as well as a running log during your flight.

Harris

	CALL SIGN	FREQ.	AS
3°	ATIS	133.8	
ALT	CPT	127.25	3
).04	GRND	121.8	9
	TWR	119.7	5
	DC	120.55	
	BOS	120.2	
ALT	NY	128.5	
).08			
	ATIS	128.15	
	AC	118.0	
)FF	TWR	119.3	
.8	GRND	121.9	

Fig. 3-7. *The sequence of frequencies for the flight to Binghamton.*

Toward the bottom of the radio frequency columns are spaces for arrival frequencies.

While working on the right side of the form, fill in the indicated airspeed (IAS) you plan to fly (top of form) and the power setting, to maintain this airspeed. Let's assume that we're making this flight to Binghamton in a Cessna 182. We enter the revolutions per minute (RPM) and manifold pressure—2,300 and 23″—that we estimate will maintain 125 knots (FIG. 3-8).

Fig. 3-8. *Performance details, flight to Binghamton.*

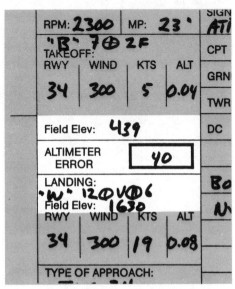

Fig. 3-9. *Field elevations, flight to Binghamton.*

FIELD ELEVATION

When still on the right side of the flight log form, enter the elevation of departure and destination airports (FIG. 3-9). Because you don't know the altitude of takeoff and landing runways you will be assigned, enter the field elevations.

In your VFR flying days, the field elevations you picked off your sectional charts were accurate enough for flying a good pattern and making a good landing. But field elevation is the highest point on an airport's usable runways, and not necessarily the elevation for an instrument approach. More precise field elevations are utilized.

Take a closer look at the approach chart for the ILS 34 approach at Binghamton (FIG. 3-4). On the airport diagram you will see "TDZE 1597" at the approach end of Runway 34. TDZE stands for "touch down zone elevation." This is the precise field elevation for the landing area of this instrument approach and it is this TDZE of 1,597 feet on which the landing minimums for runway 34 are based, not the field elevation of 1,630 feet. Keep in mind the fact that no matter which runway you end up using, the TDZE may be substantially different from the field elevation.

DATE November 30   PILOT: Sollmann/Harris

FLIGHT PLAN ROUTE

Westchester 9 Haays △
D-HUO V252 Latty △

ATC CLEARS N 3456 X

C - BGM W-9 Haays F

	TAS: 125	IAS:	
	RPM: 2300	MP: 23'	

TAKEOFF: "B" 70 @ 2F

RWY	WIND	KTS	ALT
34	300	5	0.04

Field Elev: 439

ALTIMETER ERROR: 40

LANDING: "W" 12 @ OVC6

RWY	WIND	KTS	ALT
34	300	19	0.08

Field Elev: 163e

TYPE OF APPROACH: ILS 34

CALL SIGN		FREQ.	ASGND ALT
ATIS	CPT	133.8	30
		127.25	30
GRND		121.8	90
TWR		119.7	80
DC		120.55	40
BOS		120.2	
NY			

Flight log table:

FIX NAME	FREQ	IDENT	INTERSEC NAME/IDENT	FREQ/RADIAL	ROUTE VIA	M.C. FROM	M.C. TO	DIST. REM	G/S EST	G/S ACT	TIME ENR EST	TIME ENR ACT	TIME ARR EST	TIME ARR ACT
—			Haays STVO	116.1 / 054	W-9	305	374	108		26		31		
HUO	116.1					325							33	
Hugenot					V273	325	8	144				37		
HUO	116.1					314							38	
			NY 116.9 / 272		V252	318	66	134						
BGM	112.2													
BGM airport					ILS	340	14	109						
I-BGM	110.3							125	(1…					

Fig. 3-10. Flight log form for the flight to Binghamton with all the details that can be filled in prior to the weather briefing.

The same applies to your departure airport. You will need an equally precise elevation to set your altimeter accurately. At the ramp where I park at Westchester Airport, the spot elevation is 388 feet, which is 51 feet less than the published field elevation (highest point) for Westchester!

If an altimeter had the maximum allowable instrument error of 70 feet, it could be in error a total of 121 feet if you used Westchester's official field elevation. Most precision approaches have a decision height 200 feet above the TDZE. If your altimeter was off by 121 feet, you could have serious problems on a precision approach.

If you are operating from an unfamiliar field and cannot find the elevation figure for the parking ramp, use the TDZE elevation nearest the run-up area for your takeoff runway.

AIRPORT SERVICES

This completes the information you can enter on your flight log form prior to your weather briefing (FIG. 3-10).

But some unofficial planning items you should think about can make the difference between an easy, comfortable journey and a difficult trip. Just because the airport is listed in the A/FD doesn't mean that it's adequate for you.

A lot of important information is not listed in the A/FD. Are there special noise abatement procedures? Will the FBO be open for parking, fuel, and other services when you arrive? Is there convenient transportation to the office meeting or to that golf course you've been wanting to play for so long?

Answer these questions before you depart. The best way to obtain this information is from somebody on the ground at your destination who knows the area. Determine ahead of time which FBO you'll be using and give them a call. *AOPA'S Airports USA* is an excellent current source of telephone numbers for FBOs. This annual guide also provides a wealth of information on transportation, lodging, and other services.

4
Weather/Whether to Fly?

UPON COMPLETION OF THE FIRST PHASE OF FLIGHT PLANNING, YOU MUST DETERMINE if there is anything in the current weather or forecast weather that might hinder the flight. If anything, weather considerations are simpler for an IFR flight than for a VFR flight. In IFR flying, it's assumed that you'll be in the clouds all the way until you break out during your final approach.

In VFR flying, a good part of the weather analysis is devoted to figuring out where the clouds are, whether to go above, below, or around them, and can you get to your destination before everything begins closing in? One joy of an instrument rating is setting off on an IFR cross-country when your VFR-only friends are still back at the FBO agonizing over all those VFR weather decisions.

Your decision whether or not to make an IFR flight boils down to five major go/no-go decision factors:

1. Thunderstorms.
2. Turbulence.
3. Icing.
4. Fog.
5. Departure and destination weather minimums.

THUNDERSTORMS

Thunderstorms are formed when moisture is combined with rising columns of unstable warm air. As the warm air rises it cools and condenses into droplets of rain. Heat is released during condensation, and this additional heat, in turn, increases the speed and power of the rising column of air.

As a thunderstorm begins to tower thousands of feet up into extremely cold temperatures, the rising column of warm air finally cools and a strong column of descending air begins to plunge toward the ground outside the core of rising warm air. Rain droplets carried to the top of the thunderstorm are frozen and begin to spill out of the top of the storm as hail.

This pattern is typical of the buildups we find in the summer. But we don't want to fool ourselves with the idea that thunderstorms are just a summertime problem. We can have thunderstorms in January most anywhere in the country. Anytime you have lifting, unstable air and moisture in the atmosphere, thunderstorms can develop.

This is a simplified description of a very complex process, and you can and should learn more about thunderstorms by studying weather related publications listed in Appendix A. The FAA book *Aviation Weather* has a particularly good chapter on thunderstorms and good discussions of turbulence, icing and fog.

What does all this mean to an instrument pilot? The point is that thunderstorms contain powerful columns of rising warm air surrounded by equally powerful columns of descending cold air.

The updraft in a mature thunderstorm cell may exceed 6,000 feet per minute (fpm) and passing through the shear between the updraft and downdraft structural limits of most general aviation aircraft may be exceeded. The vicinity of a thunderstorm must be avoided at all times. The odds are against you in a light plane.

Hail

Other thunderstorm dangers are not so obvious. When hail spews out of the anvil shape at the top of a thunderstorm, it can come down in clear air as far as eight or 10 miles from the storm. A good rule of thumb to avoid getting knocked out of the sky by a shaft of hail is to circumnavigate a storm by 20 miles or more.

This will also keep you outside the gust front that rings a mature thunderstorm. The gust front is an area of heavy turbulence caused by the descending currents of cold air reaching the ground and spreading outward and are commonly called downbursts. These gusting cold air currents are a cause of low-level wind shear, and they can transform an otherwise routine instrument approach into a disaster or, at best, a hostile environment you have to fight all the way.

Imbedded Thunderstorms

Of particular concern to the IFR pilot are imbedded thunderstorms. As the name implies, imbedded thunderstorms are hidden by the low-level clouds associated with frontal systems. As a front moves, its wedge-like leading edge forces columns of air aloft, creating columns of rising warm air. The moisture present in the clouds, plus the rising warm air make for an ideal thunderstorm scenario. Surrounded by clouds, the imbedded thunderstorm cannot be spotted visually, and you could penetrate one without warning.

The possibility that you may encounter an imbedded thunderstorm can be predicted by meteorologists with fair accuracy. The National Weather Service (NWS) systematically measures air stability (the lapse rate), temperature, and humidity at various altitudes, and analyzes these variables along with other factors to develop forecasts.

Thunderstorm Forecasts and Reports. If thunderstorms are forecast, this will always be stated in the terminal and area forecasts that are reported during a weather briefing.

Thunderstorms are also included in AIRMETs and SIGMETs (Airman's Meteorological Information and Significant Meteorological Information). These advisories are issued to amend Area Forecasts and to announce weather that is significant or potentially hazardous to flight. The advisories cover thunderstorms, turbulence, high winds, icing, low ceilings, and low visibilities.

AIRMETs concern weather of less severity than weather covered by SIGMETs. AIRMETs and SIGMETs will normally be part of all weather briefings. If not, ask. The news they contain—bad or good—is important to making your go/no-go decision.

Ground-based weather radar accurately pinpoints developing storms and tracks them as they grow in size and power. Air traffic control radar, on the other hand, blocks out weather to a large extent to follow air traffic more precisely. It is important to understand this distinction because air traffic controllers cannot always provide deviation information to get around thunderstorms. Thunderstorms don't always show up clearly on traffic control radar.

Data from specialized weather radar locations is plotted on radar summary charts. The information from these charts can help you steer clear of thunderstorms while planning your flight.

With centralization of Flight Service Stations (FSSs) and the introduction of extensive telephone briefing services, most pilots will obtain weather information over the telephone without seeing the printed versions of radar summary charts, surface analyses, area and terminal forecasts, hourly weather reports, and other reports and charts provided by the National Weather Service.

Nevertheless, you must know what information these different reports contain. Your knowledge should be sufficient so you can visualize what the weather briefer is reading to you over the telephone and you should know what to ask for if you need special information. There are also many questions about aviation weather services on the instrument written examination.

I recommend that you purchase a copy of the FAA's book *Aviation Weather Services* (listed in Appendix A). Study descriptions of the forecasts, reports, and charts relevant to light plane IFR flying.

Locate the central Flight Service Station (FSS) serving your area and visit it. (*Instructor Note:* File and fly an IFR cross-country to the FSS airport.) The FSS personnel will show material they use for weather briefings and discuss it. You may be surprised how much they rely on computer displays rather than paper printouts these days!

If thunderstorms are forecast for any portion of your flight—or they are developing under the watchful eye of a weather radar—choose a different route or destination or don't make that flight. With all the assistance available there is no excuse for flying into a thunderstorm, either VFR or IFR. During rapidly increasing thunderstorm activity, a Severe Weather Avoidance Plan (SWAP) may be in effect to help all aircraft fly more favorable routes.

TURBULENCE

While you can always count on getting a rough ride in turbulence in and around thunderstorms, turbulence can also occur along fast moving weather fronts and around intense high and low pressure boundaries.

For the instrument pilot, avoiding turbulence takes on another dimension. It's not just a matter of avoiding severe and extreme turbulence that may cause structural damage to your aircraft. You must also consider the effect of light and moderate turbulence on personal performance as an instrument pilot.

It is much harder to hold headings and altitudes on instruments if buffeted by turbulence. Maximum attention and a lot of work are required to maintain heading and altitude in IFR flight in turbulent conditions. Fatigue sets in much more quickly than on a smooth flight.

Fatigue is not just uncomfortable, it is a very real danger. As the AIM points out in its section on "Medical Facts for Pilots": "Fatigue continues to be one of the most treacherous hazards to flight safety, as it may not be apparent to a pilot until serious errors are made."

And it's not just in the clouds that turbulence is a problem, but in clear air as well. If you have to contend with turbulence over a long stretch during the VFR portion of an IFR flight, you will be fatigued, tense, and less confident when descending into the clouds to make an instrument approach. Tired, tense, and anxious is not the best way to commence a good instrument approach.

High Winds

Turbulence is also associated with high winds. Expect light turbulence when winds of 15 to 25 knots are forecast or reported; moderate turbulence in winds of 25 to 50 knots. Base the go/no-go decision on whether or not you can handle winds of up to 50 knots and the associated turbulence. Above 50 knots lies the realm of severe and extreme turbulence. Turbulence in this range can cause structural damage to a light plane, as well as slam you around unmercifully. Don't try it!

If there is a possibility of light or moderate turbulence during the flight—and particularly if it looks like the turbulence might last for a long time—decide if you will be able to handle an instrument approach in lousy weather conditions and keep headings and altitudes under control. If there is any doubt, plan a shorter flight, pick a destination for a VFR approach and landing, or wait for another day.

Terminal and area forecasts both contain wind information and the area forecast specifically covers turbulence. AIRMETs and SIGMETs cover both high winds and turbulence.

A further aid is the surface weather analysis chart that you became so familiar with during your VFR training. If the isobars are tightly packed with little space between them you can expect high winds and turbulence.

A pilot weather report (PIREP) is even more conclusive evidence of turbulence. If the pilot of an aircraft similar to yours reports turbulence along the route, take it seriously! Here is someone who has just been through a troublesome or even hazardous experience and is telling it like it is—something no one in a weather station on the ground can ever do.

If a pilot reports turbulence along the route of flight, pick a different route or a different altitude or cancel your flight if you don't think you can handle it. If you are airborne, don't hesitate to ask your controller to solicit PIREPs on turbulence (or icing or any other conditions that might affect your flight). Likewise, if you encounter icing, you should report it immediately. Your anonymous fellow airmen flying behind you will be grateful to you.

ICING

The situation is different with icing conditions, because it is impossible to forecast icing with certainty. But conditions that produce potential icing are well known and if these conditions are present you might well assume that you will pick up a load of ice on the airframe, so make other plans.

Aircraft structural icing requires moisture and below-freezing temperatures. Ice may form on your aircraft if you fly in clouds above the freezing level. if the freezing level is at or near the ground, you may experience icing as soon as you enter the clouds after takeoff.

Consider this for a moment. Takeoff and departure are among the busiest and most intense moments of an IFR flight. You are making the transition from visual to instrument flight—sometimes very quickly if your departure ceiling and visibility are low. You are switching from tower to departure control and departure control may well have a clearance amendment for you. You may not be aware of it, but every sense is alert to abnormal sounds from the engine, unusual control pressures, and unusual instrument readings—is this any time to be worrying about ice?

The most dangerous form of icing occurs when a mass of rainy, warm air overlies or overrides a cold air mass. As the warmer rain falls through the lower, colder air mass the rain becomes cold enough to freeze on impact with the surfaces of the aircraft. The cold droplets hit, splatter, and coat the surfaces they strike with layers of ice that can build up rapidly.

This rapid build-up can add hundreds of pounds of weight to a small aircraft very quickly. The ice destroys aerodynamic characteristics of wings, control surfaces, and propellers. In extreme cases, two to three inches of ice can form on the leading edge of an airfoil in less than five minutes. It takes only ½-in. of ice to reduce the lifting power of some aircraft by 50 percent. No doubt about it—icing can be lethal.

Freezing Level

If your flight is going to be in the clouds, plan to stay below the freezing level and avoid areas where warm, moist air overlies colder air. Where do you get information on the freezing level and icing? The area forecast is the first place to look. The National Weather Service "sounds" the atmosphere for temperatures at hundreds of locations throughout the United States and plots freezing levels from these soundings. If the data indicates the possibility of icing, it will be included in an area forecast in a special "Icing and Freezing Level" section.

Once again the most reliable information comes from PIREPs. If a pilot reports icing along your route of flight, pick a different route or cancel your flight. The pilot isn't guessing or forecasting about this hazard; he can actually see the ice building up on the aircraft. This is much more useful than a forecast, which can only predict conditions favorable for icing.

FOG

If you are on the ground watching a wall of fog move in, it seems to have a sinister, menacing appearance. And with good reason. In a matter of minutes, fog can reduce visibility to zero. In a thick "pea soup" that you might encounter along the coast of New England, fog can be so dense that you literally cannot see where you are walking. This can be very bad news for pilots, particularly if you are not prepared for it and the fog comes as a surprise.

The best preparation is to understand the conditions that produce fog and then avoid those conditions. Fortunately, this is not very complicated. Fog forms either when air is cooled to its dew point or when the dew point is raised by the presence of additional moisture. Raising the dew point is usually accomplished by the evaporation of water from falling precipitation or by the passage of a body of air over a wet surface.

Fog classifications are: radiation, advection, upslope, precipitation-induced, and ice. You should deepen your understanding of this phenomenon through further study of the related weather publications listed at the back of this book.

Dew Point Spread

Theory aside, dew point is the temperature at which air is saturated and the water vapor begins to condense and produce visible moisture; dew point is expressed as a temperature. If the air temperature drops to the dew point temperature, fog forms. This is the most common situation. Along coastal areas, however, moist air moving in from the sea can raise the dew point and produce fog.

Because fog forms so low to the ground, there is nothing like it for closing an instrument runway or an entire airport. Fog can form rapidly, making it especially hazardous. You can start an instrument flight with sufficient ceiling and visibility at the destination to make a comfortable approach, then arrive and find a blanket of fog that makes it impossible to land.

Hourly aviation weather reports (SAs) will always indicate the presence of fog. The hourly reports also contain both temperature and dew point. The difference between the two numbers is the temperature-dew point spread, or simply "the spread" as it is commonly called.

When the spread is about 5° F., be alert for fog. For example, if the temperature is 48° F. in the most recent report for your destination airport, and the dew point is 43°, be prepared for fog when you arrive. The possibility of fog also appears in terminal forecasts.

If your destination is socked in with fog, there is no point in going until the fog begins to lift. If the spread is narrowing, according to reports, and drops to 5° F. or less, again there is no point in leaving until the situation begins to improve.

LOW CEILINGS AND VISIBILITY

Fog is only one of many reasons for low ceilings and visibilities. Other factors are rain, snow, haze, smoke, low pressure areas, and frontal systems. You have to ask yourself whether or not you will be able to land upon arrival. You must determine if you need an alternate airport and what the alternate might be. And there is a third item relating to minimums that many pilots overlook: Will you have sufficient ceiling and visibility at your *departure* airport to return for an immediate landing if necessary?

Let's take this third item first. Basically, under FAR 91, you can take off in any weather—including zero-zero—if you're not carrying passengers or cargo for hire or pay. (See FAR 91.116 - "Takeoff and Landing Under IFR," paragraph (f) in Appendix B.)

If you read this regulation closely, notice that takeoff minimums apply to FAR Part 121, 123, 125, 129, and 135 operations only, but Part 91 is not mentioned. Therefore Part 91 operations are not covered by this regulation.

However, a takeoff in conditions approaching zero-zero is certainly not very smart. What do you do if a door pops open or the engine doesn't sound right? Where can you land?

The "Departure Alternate"

Two answers to this question should always be part of your flight planning. One, do not depart on an instrument flight if the ceiling and visibility at your departure airport are too low for a safe return in an emergency, possibly an instrument approach.

Or two, if the ceiling and visibility at departure are too low to return, select a departure alternate—a nearby airport with adequate ceiling and visibility for a safe arrival. Establish a route to that field, study the ATC frequencies and details of the approach like you would for the destination. In other words, be completely prepared to land at this departure alternate if necessary.

If departing from an airport that has no instrument approach, select solution number two. Pick out the nearest field with an instrument approach plus ceiling and visibility above your "personal" minimums and be prepared to make that approach. It's certainly much smarter to have an "out," or alternative when taking off under IFR conditions, no matter what type of airport you are departing from.

DESTINATION MINIMUMS

After analyzing your departure situation, you must next anticipate the kind of weather you're going to have at your destination airport. Now let's take a closer look at the meaning of minimums. You are familiar with ceilings and visibilities from VFR training; in IFR flying they take on additional importance.

The *Airman's Information Manual* (AIM), published by the FAA, defines ceiling and visibility in the "Pilot/Controller Glossary." Appendix C contains pertinent IFR glossary. However, you should take time to browse through all definitions in the AIM glossary. it contains a gold mine of information relating to IFR operations, and each nugget is as clear and concise as anything you will ever find.

Ceiling is: The heights above the earth's surface of the lowest layer of clouds or obscuring phenomena that is reported as broken, overcast, or obscuration, and not classified as thin or partial.

VISIBILITY

The AIM Glossary lists and defines several visibility classifications in increasing order of precision.

Ground Visibility. "Prevailing horizontal visibility near the earth's surface as reported by the National Weather Service or an accredited observer."

Prevailing Visibility. "The greatest horizontal visibility equalled or exceeded throughout at least half the horizon circle which need not necessarily be continuous."

Runway Visibility Value/RVV. "The visibility determined for a particular runway by a transmissometer. RVV is used in lieu of prevailing visibility in determining minimums for a particular runway."

Runway Visual Range/RVR. "An instrumentally derived value . . . that represents the horizontal distance a pilot will see down the runway from the approach end . . . RVR, in contrast to prevailing or RVV, is based on what a pilot in a moving aircraft should see looking down the runway. RVR is used in lieu of RVV and/or prevailing visibility in determining minimums for a particular runway."

Which definition is used when flight planning?

Use the most precise visibility measurement available. Weather reports for airports with ILS approaches will contain both a ground visibility and an RVV or RVR for the runway in use. Visibilities at airports with ADF, VOR, or the non-precision approaches will usually be given in the less precise prevailing visibility.

Importance of the Visibility Minimum

Why does the FAA go to such great lengths to define visibility?

The answer is quite clear: "No pilot operating an aircraft . . . may land that aircraft when the flight visibility is less than the visibility prescribed in the standard instrument approach procedure being used." This is the law as stated in FAR 91.116 (d) which you will find in Appendix B. In other words, you cannot land if the visibility is below minimums.

Furthermore, according to another section of the same FAR, you cannot descend below a minimum altitude if the visibility prescribed for your approach is less than minimums. There is a good reason for this because it is possible to be clear of clouds upon reaching altitude minimums, but you may be unable to see far enough down the landing runway to make a safe approach because of fog, rain, snow, or other runway-obscuring condition.

Understand from the beginning that *visibility*, not ceiling, determines whether or not you can initiate an instrument approach.

Let's continue planning our flight to Binghamton, New York. What ceiling and visibility minimums apply to this flight? Turn first to the approach chart for ILS RWY 34 to Binghamton and look at the landing minimums section (FIG. 4-1).

Approach Categories

Notice Categories A, B, C, and D. These are "Aircraft Approach Categories based on actual final approach speed. Category A is for approach speeds from 0 (for helicopters) to 90 knots. Unless you are training for an instrument rating in a Lear Jet or something equally exotic, Category A will be used on all training flights. (For further information on aircraft approach categories, see the inside front cover of any NOS approach chart booklet.)

Because you plan to make the ILS 34 approach, use the minimums listed for the straight-in approach (S-ILS 34): 1797/24. The numbers after it indicate:

200 - height above touchdown (HAT) at the ceiling minimum.
200-½ - ceiling and RVR converted to prevailing visibility in statute miles.

The ceiling and visibility minimums for planning this flight are the third set of numbers: 200 feet and one-half mile. Now check the other approach charts for Binghamton and see if any have lower minimums; shifting winds may change arrival runways. There are

Fig. 4-1. *Landing minimums for Binghamton ILS RWY 34 approach, as shown on the NOS approach chart.*

CATEGORY	A	B	C	D
S-ILS 34	1797/24 200 (200- ½)			
S-LOC 34	1960/24 363 (400-½)			1960/40 363 (400-¾)
CIRCLING	2060-1 430 (500-1)	2080-1 450 (500-1)	2080-1½ 450 (500-1½)	2180-2 550 (600-2)
Cat. D S-LOC visibility increased to RVR 5000 for inoperative MM.				

five instrument approaches to Binghamton. No minimums are lower than those for ILS 34, although the minimums for ILS 16 are close at 300 feet and one-half mile.

To make a ''go/no-go'' decision about this flight, look for a ceiling of no less than 200 feet and a visibility of no less than one-half mile. If ceiling and visibility are forecast to be lower than this at your estimated time of arrival (ETA), you will probably not be able to land. It is perfectly legal to file IFR and fly to Binghamton and attempt an approach when the weather is forecast to be below minimums (air carrier pilots are not allowed to do this), but what's the point of making the trip if you can't land?

Well, you might go to Binghamton and try an approach just to see if the weather had improved enough to land. You might get lucky! On the other hand, the weather could be much worse upon arrival and a landing might be impossible. This can happen on even the best-planned flights, especially in winter when weather systems pick up speed and sweep across the country much faster than the summer.

You must pick an alternate airport for a safe landing if you can't get into Binghamton. This is not just a good idea, it's required by FAR 91.83(a). You may have to pick an alternate even if Binghamton is forecast to be VFR upon arrival. Many people are fooled by this because they think ''The airport is going to be VFR when I get there so I don't need an alternate.'' That's not true. To file to an airport IFR without listing an alternate, you need twice the ceiling required for VFR.

THE ''ONE, TWO, THREE RULE''

If the ceiling and visibility are forecast for one hour before ETA through one hour after ETA to be 2,000 feet and three miles, according to FAR 91.83 (b) (Appendix B), a pilot must select an alternate airport and include the selection on the flight plan. That is the ''one, two, three rule''—one hour, two thousand feet, and three miles.

Another rule of thumb to help remember the alternate airport requirements is: You must have VFR conditions plus 1,000 feet for ETA $\pm$ one hour or you must file an alternate.

SELECTING AN ALTERNATE

How do you pick an alternate airport? Obviously, you can't pick an alternate that is so far away you'll run out of fuel before arrival. And there's not much point in picking an alternate where the weather is so bad it prevents landing.

The FARs are grounded in good common sense on these two points. First, FAR 91.23 (Appendix B) says you must carry enough fuel on an IFR flight to: (1) complete the flight to the first airport of intended landing; (2) fly from that airport to the alternate, if one is required, and (3) fly after that for 45 minutes at normal cruising speed.

You can list an airport as an alternate only if the ceiling and visibility forecast for the alternate at your time of arrival will be at or above 600 feet and two miles (if the airport has a precision approach), or 800 feet and two miles (if it has only an ADF, VOR, or other nonprecision approach).

There may be a catch to this, however. Some airports may not be authorized for use as alternates, while others available for alternates may have *higher* minimum requirements than 600/2 and 800/2 because of local conditions such as hills, towers, or radio towers. How can you find this out?

Turn to the front of the NOS approach chart booklet and find a listing of "IFR Alternate Minimums." Whenever an airport has minimums that deviate from the standard 600/2 and 800/2, it will be listed in this section (FIG. 4-2). Note that an airport such as Farmingdale Republic is not allowed for use as an alternate when the control tower is not in operation.

If the airport selected as an alternate is listed in this section, use the minimums in this section. If the choice of an alternate does not appear in this section, use the 600/2 and 800/2 minimums discussed above.

Jeppesen approach charts list the alternate airport minimums on the back of the first approach chart for each airport, regardless of whether or not the minimums are standard or nonstandard (FIG. 4-3).

PERSONAL MINIMUMS

Just as most airlines qualify their crews to fly certain minimums, many competent instrument pilots set higher minimums for themselves than the published minimums. This is very smart! A brand new instrument pilot might want to start out with a ceiling of 1,000 and three miles visibility (VFR minimums) until gaining more experience and confidence. This can then be lowered, depending upon frequency of your flights until reaching the lowest minimums available.

If you don't fly very often and are barely meeting currency requirements, it might be prudent to raise personal limits to 500 and 2, for example. Many pilots make it a policy to fly instruments every three or four weeks. Many also sign up for instrument refresher instruction every three months if they feel a little rusty.

To maintain currency as an instrument pilot, regulations require six hours of instrument flight and six instrument approaches each six months. Just because you are legally current does that mean that you are competent to fly the published minimums? *Absolutely not!* Set personal minimums that you are comfortable with and feel safe.

SUMMARY OF WEATHER FACTORS

Let's pause to summarize the necessary weather information when you request a weather briefing:

1. Are there any weather conditions that might make it difficult or impossible to complete the flight as planned, such as:
   - Thunderstorms
   - Turbulence
   - Icing
   - Fog

⚠   ⚠

INSTRUMENT APPROACH PROCEDURES (CHARTS)

⚠ IFR ALTERNATE MINIMUMS
(NOT APPLICABLE TO USA/USN/USAF)

Standard alternate minimums for nonprecision approaches are 800-2 (NDB, VOR, LOC, TACAN, LDA, VORTAC, VOR/DME or ASR); for precision approaches 600-2 (ILS or PAR). Airports within this geographical area that require alternate minimums other than standard or alternate minimums with restrictions are listed below. NA - means alternate minimums are not authorized due to unmonitored facility or absence of weather reporting service. Civil pilots see FAR 91. USA/USN/USAF pilots refer to appropriate regulations.

NAME	ALTERNATE MINIMUMS	NAME	ALTERNATE MINIMUMS

ADIRONDACK — SEE SARANAC LAKE, NY

ALBANY COUNTY, NY
ILS Rwy 1*
VOR Rwy 1**
VOR Rwy 19†
RADAR-1†
*Category D, 800-2½
**Category C, 800-2¼, Category D, 800-2½
†Category D, 800-2¼

ALLENTOWN-BETHLEHEM-EASTON, PA
ILS Rwy 13
ILS Categories A, B, C, 700-2; Category D, 700-2¼; LOC Category D, 800-2¼.

ALTOONA-BLAIR COUNTY, PA
ILS Rwy 20*
VOR-A†
*ILS, LOC, Categories A, B, C, 900-2½, Category D, 1100-3.
†Category D, 1100-3

CAPITAL CITY — SEE HARRISBURG, PA

CHAUTAUQUA COUNTY — SEE JAMESTOWN, NY

CHESS-LAMBERTON — SEE FRANKLIN, PA

DUTCHESS COUNTY — SEE POUGHKEEPSIE, NY

ELMIRA/CORNING REGIONAL, NY
NDB Rwy 24, 1200-3
ILS Rwy 6†↓
ILS Rwy 24**
**ILS, LOC, 1200-3
†ILS, Categories A, B, 1200-2, Categories C, D, 1200-3; LOC, Categories A, B, 1200-2, Categories C, D, 1200-3.
↓NA when control tower closed.

ENDICOTT, NY
TRI-CITIES . . . . VOR-1 NA

ERIE INTL, PA
ILS Rwy 6, 700-2*
ILS Rwy 24, 700-2*
NDB Rwy 24
RADAR-1
NA when control tower closed.

FARMINGDALE, NY
REPUBLIC . . . . LOC (BC) Rwy 32
ILS Rwy 14
NDB Rwy 1
NA when control zone not effective.

FRANKLIN, PA
CHESS-LAMBERTON . . . . ILS Rwy 20
VOR Rwy 2
VOR Rwy 20
NA when control zone not in effect.

GREATER PITTSBURGH — SEE PITTSBURGH, PA

HARRISBURG, PA
CAPITAL CITY . . . . ILS Rwy 8*
RADAR-1, 1100-3
*ILS, 700-2

HARRISBURG INTL AIRPORT-OLMSTED FIELD — SEE MIDDLETOWN, PA

ISLIP, NY
LONG ISLAND MACARTHUR . . . . ILS Rwy 24
LOC, NA.

NE-2

⚠   ⚠

Fig. 4-2. If IFR landing minimums are other than 600-2 for precision approaches and 800-2 for nonprecision approaches, they will be listed in this "IFR Alternate Minimums" section of the NOS Instrument Approach Procedures booklet.

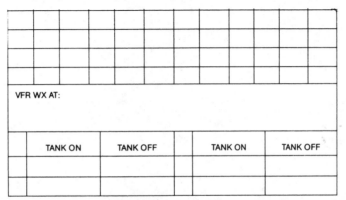

FOR FILING AS ALTERNATE	
Precision	Non-Precision
600-2	800-2

A
B
C
D

Fig. 4-3. *On a Jeppesen approach chart, an airport cannot be used as an alternate unless the ceiling and visibility at arrival time will be at or better than stated in the "For Filing as Alternate" block on the back of the first chart for the airport.* Reproduced with permission of Jeppesen Sanderson, Inc.

VFR WX AT:

TANK ON	TANK OFF		TANK ON	TANK OFF

Fig. 4-4. *Space for noting nearest VFR weather on flight log form.*

2. Are the ceilings and visibilities high enough at your departure and your destination airports for a safe IFR takeoff and for a safe (and legal) IFR approach and landing? Does the weather meet your personal minimums?

3. Do you need an alternate? If so, what alternate airports can you reach that meet the minimum requirements for ceiling and visibility?

Look again at the flight log form and notice in the lower right-hand corner the words "VFR WX AT:" (FIG. 4-4). This item is not found in any FAA publications; it is based on years of IFR flying and IFR instruction.

When the chips are down, where you can find VFR conditions? Where can you go for a safe VFR landing if all electrical power—radios, transponder, electrically-powered instruments, pitot heat, lights—is lost? Where can you go if an instrument flight becomes horrendous?

Always learn where the nearest VFR conditions are to safely abort the IFR flight and land VFR. A weather briefer will provide this information if requested. If there is no VFR weather at an airport within range, don't go!

"When in doubt, wait it out!"

5
How to Get a Good
Weather Briefing

IT WASN'T TOO LONG AGO THAT GETTING THE WEATHER MEANT TALKING DIRECTLY TO A weather briefer, either by telephone or during a personal visit to a Flight Service Station (FSS). Pilots lucky enough to have an FSS located at their airport might still go in and examine all the weather maps, forecasts, and reports, then talk to a weather briefer in person.

But for most pilots this personal contact is a thing of the past. The FAA has embarked on an ambitious program for centralizing and automating FSS facilities.

"(A better) relationship between pilot and briefer was possible during an earlier period, when the FAA had over 300 flight service stations and half the number of flight operations undertaken today," says the AOPA Air Safety Foundation's *Flight Instructors' Safety Report*.

"That time is gone—history. By the close of the 1980s most of the individual stations will have been phased out in favor of 61 Automated Flight Service Stations (AFSSs), which will provide at least twice the service that has been available from all the non-automated stations."

AUTOMATED FLIGHT SERVICE

Like it or not, the highly automated FSS with its recorded weather information is the wave of the future. The bottom line, as always, is money. The FAA simply does not have the budget to operate the number of Flight Service Stations it once had, nor can it justify the cost of extensive personal briefings for everyone, as was once the case.

You may still get a "live" weather briefing for every cross-country flight, VFR or IFR, by talking directly to an FSS specialist to assure a complete briefing on NOTAMs, PIREPs, and fast-changing conditions affecting a specific flight plan.

But a live weather briefing will be much shorter and more to the point. And the specialist may assume that the pilot has already availed himself of the many other sources of weather information before the briefing.

The new automated systems can save time and effort—and make planning simpler—if a pilot understands what to expect from the FSS specialist and what is expected from the pilot.

It is much easier these days to get good accurate weather information than ever before. Some information will come from non-standard sources.

TELEVISION WEATHER

Television weather broadcasts during the local evening news are an excellent source of information on weather trends. With the use of satellite photographs and computer-generated displays, these programs have become sophisticated in recent years. The information TV weather programs provide on fronts, pressure systems, precipitation, and temperatures show "the big picture" of what the next day's weather will be.

Begin the weather briefing the evening before a flight with a look at the TV weather. Look for the big patterns—fronts, pressure systems, and bands of clouds in the satellite photographs—and ask what effect these large patterns will have on the flight.

Is the weather improving or deteriorating? Are low ceilings and visibilities a possibility at departure and destination airports? Will the flight be in the clouds or in the clear? Is there a chance of thunderstorms, turbulence, icing? Don't look for precise details available from aviation weather sources. Look instead for trends that can be verified in the detailed weather briefing.

On the morning of a flight check the TV weather again to see how conditions changed overnight. Cross-check with newspaper weather reports.

When *USA Today* introduced its big, full-color, full-page "Weather Across the USA" feature several years ago, many other newspapers spruced up their weather maps to stay competitive. So newspaper weather coverage is even better.

If you have cable TV with "The Weather Channel," you can get a minute-by-minute update of both local and national weather. This is an excellent resource for pilots, because much of the weather information comes from National Weather Service weather reports and forecasts, the same source for FSS weather briefings.

"A.M. Weather"

Another excellent TV weather service for pilots is "A.M. Weather." This award-winning national weather program is the next best thing to a skilled aviation weather briefer. These 15-minute, early morning programs are carried by almost 300 TV stations from coast to coast.

Some professional pilots make a habit of watching this program *every day*, regardless of flight schedules. For instrument students, "A.M. Weather" does more than just give a good picture of the day's weather; it will also improve an understanding of weather because the commentators make a point of discussing how and why weather phenomena and patterns develop. It is a learning experience on weather.

TWEB

Commercial radio stations can also be excellent sources of weather information. While some commercial stations do a good job of weather reporting, the best bet is tuning into a Transcribed Weather Broadcast (TWEB).

The information in TWEBs is more specific than TV and other radio broadcasts. In fact, you can sometimes get almost all details necessary to plan a flight from a TWEB, including hourly weather reports about departure and destination airports, area forecasts covering the route, and winds aloft reports to compute ground speed.

Selected NDBs and VORs carry TWEBs and are indicated in the *Airport/Facility Directory* (AFD) with this symbol: Early in instrument training, a pilot should examine the "green book," locate the TWEBs within 300 to 400 miles and list them.

TWEBs in the low-frequency (LF) range can occasionally be heard in the 200-400 band on the ADF receiver. But this is the exception rather than the rule. Throughout New York state, where I am based, there are only six TWEBs, and only two originate from NDBs in the low frequency range; other TWEBs are broadcast by VOR stations.

Because this is the case I strongly recommend investment in a portable receiver that will pick up the aviation Very High Frequency (VHF) band (108-136 MHz) and the aviation Low Frequency (LF) band (200-400 KHz) to hear TWEBs and ATIS broadcasts at home or at the airport. Portable receiver prices start at less than $100 in aviation supply outlets.

While you're at it, check out the prices and options on battery-powered VHF transceivers. For a bit more money a transceiver will provide emergency communications capability in flight, as well as receive broadcasts on the ground. It could save the day during a "lost communications" emergency in IFR weather. A portable transceiver may also be used for receiving a clearance from ATC prior to engine start.

TWEBs are also one option of the toll-free 800 service established by the FAA. Call 1-800-WX-BRIEF (1-800-992-7433) from almost any telephone in the United States, as well as many foreign exchanges, for automatic routing to the appropriate FSS.

PATWAS

This is not a radical development but merely an improvement and extension of the older PATWAS (Pilots' Automatic Telephone Answering Service). In fact many pilots—and even the FAA—refer to the 1-800-WX-BRIEF service as PATWAS.

FIGURE 5-1 is a summary of services available through 1-800-WX-BRIEF for New York Automated FSS. This FSS serves the area around Westchester County Airport, where

I am based. Follow the instructions and key in the touch-tone numbers for the respective item on the menu for an automatic connection to that item.

If touch-tone service is unavailable, a flight service specialist will come on the line. Tell him the name or number of the desired item or items for faster service.

Recorded Briefings

This is the transcript at New York Automated FSS through 1-800-WX-BRIEF:

"Thank you for calling. For faster service, please press or dial one, now." (Press or dial 1, and the recorded message continues.)

"Welcome to the New York Automated Flight Service Station. Users without touch-tone telephones please remain on the line and you will be answered by the next available pilot weather briefer.

"Users with touch-tone telephones, after the ring, enter the appropriate three-digit code for the service desired or dial two-zero-zero for the main menu.

"It may be helpful to have a pencil and paper handy."

200	Main menu of services.
300	Recorded weather and PATWAS menu.
301	Calverton VOR TWEB.
302	Adverse weather conditions for northeastern U.S.
303	Synopsis for northeastern U.S.
304	Summary of current weather and surface winds for New York City and Long Island.
305	Summary of current weather and surface winds for the Hudson Valley.
306	Summary of current weather and surface winds for the Albany area.
307	Summary of current weather and surface winds for the route Albany to Syracuse.
308	Summary of the current weather and surface winds for the route Albany to Boston.
309	Summary of the current weather and surface winds for the route New York City to Atlantic City.
310	Forecast for New York City and Long Island.
311	Forecast for Hudson Valley.
312	Forecast for Albany area.
313	Forecast for the route Albany to Syracuse.
314	Forecast for the route Albany to Boston.
315	Forecast for the route New York City to Atlantic City.
316	Winds aloft forecast for 3, 6, and 9 thousand feet for Kennedy, Albany, Syracuse, Boston, and Atlantic City.
401	Fast file for IFR and VFR flight plans where the departure time is one hour or less from the current time, and for cancellation or amending flight plans already filed. (Rotary users: Ask for fast file one.)
402	Fast file for flight plans where the departure time is one (1) hour or more from the current time. (Rotary users: Ask for fast file two.)

Fig. 5-1. *Menu of services available through the New York Automated Flight Service Station.*

Automated FSS "main menus" vary considerably throughout the country because of local conditions and requirements; write the main menu of the FSS serving your area.

Don't worry if it takes you several tries to copy all the menu items accurately! They do speak rapidly, but it's their nickel and you can re-access the recording until all details are copied correctly.

When accessing an unfamiliar AFSS, like on the return leg of an IFR cross-country, dial or press 1-800-WX-BRIEF, then get the "main menu" by pressing 200 or requesting it. Be ready to jot down the codes for the recorded messages relevant to the flight.

Regardless of location, the sequence is the same: TWEBs, adverse weather for the area (AIRMETs and SIGMETs), synopsis for the area, current weather for several routes, forecasts for these routes, winds aloft, and, finally, the "fast file" numbers.

WEATHER BRIEFING CARD

Make your own "briefing card" that can fit onto a knee-pad or tuck into the pages of approach charts or the AFD. The briefing card should contain:

- TWEBs in the local area and respective frequencies.
- Main menu of services available through the nearest automated FSS with the codes.

(Additional card items will be introduced in the text.)

One problem with the new system is that it can often take a long time to review all the recorded messages that pertain to a flight and then file the flight plan, especially without a touch-tone telephone.

FORECAST RELIABILITY

When planning an instrument flight always ask: How good are the weather forecasts?

Pilots should understand the limitations and capabilities of present-day weather forecasting. The meteorologist understands certain atmospheric behaviors and has watched the behavior long enough to know that his knowledge of the atmosphere certainly is not complete.

Pilots who understand limitations of observations and forecasts usually make the most effective use of forecasts. The safe pilot continually views aviation with an open mind; he knows that weather is always changing and consequently knows that the older the forecast, the greater the chance that parts of it will be wrong. The weather-wise pilot looks upon a forecast as professional advice rather than an absolute surety. To have complete faith in weather forecasts is almost as bad as having no faith at all.

According to FAA summaries of recent forecast studies, pilots should consider:

1. Up to 12 hours—and even beyond—a forecast of good weather (ceiling 3,000 feet or more, and visibility three miles or more) is more likely to be correct than a forecast of conditions below 1,000 feet or less than one mile.

2. If poor weather is forecast to occur within three to four hours, the probability of occurrence is better than 80 percent.

3. Forecasts of poor flying conditions during the first few hours of the forecast period are most reliable when there is a distinct weather system, such as a front, or a trough is present. However, there is a general tendency to forecast too little bad weather in such circumstances.

4. Weather associated with fast-moving cold fronts and squall lines is the most difficult to forecast accurately.

5. Errors occur when attempts are made to forecast a *specific* time that bad weather will occur. Errors are made less frequently when forecasting that bad weather will occur during some *period* of time.

6. Surface visibility is more difficult to forecast than ceiling height. Visibility in snow is the most difficult of all visibility forecasts.

Predictable Changes

According to FAA studies, forecasters can predict the following at least 75 percent of the time:

- Passage of fast-moving cold fronts or squall lines within plus or minus two hours, as much as 10 hours in advance.
- Passage of warm fronts or slow-moving cold fronts within plus or minus five hours, up to 12 hours in advance.
- Rapid lowering of ceilings below 1,000 feet in pre-warm front conditions within plus or minus 200 feet and within plus or minus four hours.
- Onset of a thunderstorm one to two hours in advance, providing radar is available.
- Time rain or snow will begin, within plus or minus five hours.

Unpredictable Changes

Forecasters cannot predict the following with an accuracy that satisfies present aviation operational requirements:

- Time freezing rain will begin.
- Location and occurrence of severe or extreme turbulence.
- Location and occurrence of heavy icing.
- Location of the initial occurrence of a tornado.
- Ceilings of 100 or zero before they exist.
- Onset of a thunderstorm that has not formed.
- Position of a hurricane center to closer than 80 miles for more than 24 hours in advance.

With these sobering thoughts from the FAA about forecast limitations, it is time to call FSS for a weather briefing regarding the flight from Westchester County to Binghamton, New York.

It is a typical IFR day in late November in the Northeast. TV weather reports and a newspaper weather map in the morning show (FIG. 5-2) that a low pressure system with plentiful rain has been moving northeastward up the Atlantic coast. It has been raining off and on at Westchester County Airport and we can see from personal observation that the ceiling and visibility are low at the airport.

A go/no-go decision must be made. What weather elements should be considered in making the go/no-go decision? Let's review briefly the items we have already covered:

1. Thunderstorms
2. Turbulence
3. Icing
4. Fog
5. Current and forecast weather at departure and destination airports

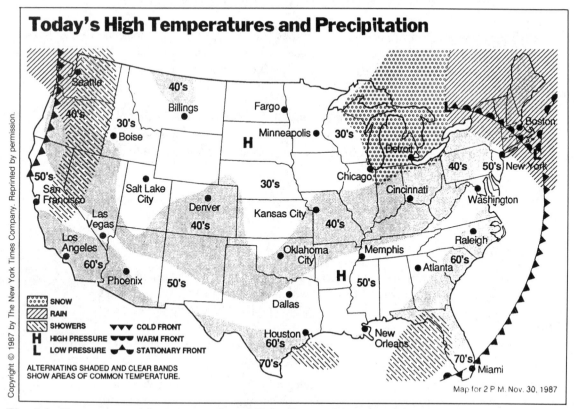

Fig. 5-2. *Newspaper weather map on day of flight discussed in text.*

6. Availability of alternate airports
7. Nearest VFR
8. Personal minimums

Now let's add two items:

9. Forecast winds aloft
10. NOTAMs

Winds aloft are necessary to determine the estimated time en route (ETE) for each leg of the flight and the total time (TT) en route, just as you did during your VFR flight planning.

NOTAMs are a critical item on all flights and are important factors when making the IFR go/no-go decision. It doesn't take much effort to imagine what you might go through at the end of a long, tiring IFR flight if you suddenly discovered that a key component of the best instrument approach was "out of service!"

FASTER SERVICE

Faster and better service is available if you initially tell the briefer:

1. The N number, which immediately identifies you as a pilot, not just someone from the general public calling to find out what the weather is like.
2. The type of airplane you are flying; light single-engine, high performance multi-engine, and jet airplanes present different briefing problems.
3. "Planning an IFR flight to (destination)."
4. Your estimated departure time (in UTC—Universal Time Coordinated).
5. Whether or not you can go IFR. The briefer doesn't know anything about you or your plane when you call and needs to know whether to provide an IFR briefing or a VFR-only briefing.

The briefer will call up the information for the flight on a computer display and will proceed step-by-step through a "standard briefing" as appropriate for your flight.

TRANSCRIBING THE WEATHER

A fancy form is unnecessary for a weather briefing. Simply list the categories of information in the proper sequence. The trouble with pre-printed weather briefing forms is they leave too much room for unnecessary information and too little room for necessary information.

Below is the sequence; items may be omitted if they are not factors in the proposed flight.

1. **Adverse Conditions.** Significant meteorological information that might influence an alteration of the proposed flight: hazardous weather, runway closures, VOR and ILS outages, and the like.

2. **Synopsis.** A brief statement of the type, location, and movement of weather systems, such as fronts and high and low pressure areas, that might affect the proposed flight.
3. **Current Conditions.** A summary of reported weather conditions applicable to the flight from surface weather reports, PIREPs, and the like.
4. **En Route Forecasts.** Summarize forecasts in logical order: departure, climbout, en route, descent.
5. **Destination Forecasts.** Note the destination's expected weather plus significant changes before and after the estimated time of arrival (ETA).
6. **Winds Aloft.** Summarize forecast winds for the proposed route.
7. **NOTAMS.**
8. **ATC Delays.** Note any known ATC delays and flow control advisories that might affect the proposed flight.

1. ADVERSE CONDITIONS
2. SYNOPSIS
3. CURRENT CONDITIONS
4. ENROUTE FORECASTS
5. DESTINATION FORECASTS
6. WINDS ALOFT
7. NOTAMS
8. ATC DELAYS

Fig. 5-3. *Quick and simple way to set up a page for jotting down a weather briefing.*

A simple way to set this up is list the eight items on the left-hand margin of a blank sheet of paper as a reminder (FIG. 5-3) and leave the rest of the page for notes. Photocopy a supply of lists with the sequence to save time during a weather briefing, or add the eight items to the briefing card as a reminder.

Weather Shorthand

Copying a weather briefing verbatim is unnecessary but I strongly suggest recording the vital highlights. FIGURE 5-4 is a short list of easy-to-use "shorthand" weather symbols and letters based on coding used by the National Weather Service. You'll remember what they mean when encountered on various weather reports, forecasts, and charts. Eventually a personal shorthand will develop. Don't hesitate to ask the briefer to read the weather slowly, especially in the beginning when the briefing form and the shorthand may be unfamiliar. Plan on having two or three blank forms at hand to write big and still get everything in.

1. **Cloud cover**

○	CLR	clear, or
◔	SCT	scattered, or
◑	BKN	broken, or
⊕	OVC	overcast, or
	X or OBS	ceiling obscured
	W or INDEF	indefinite ceiling
	E	estimated ceiling
	M	measured ceiling

2. **Weather designators**

A	hail
BD	blowing dust
BN	blowing sand
BS	blowing snow
BY	blowing spray
D	dust
F	fog
GF	ground fog
H	haze
IC	crystals
IF	ice fog
IP	ice pullets

Fig. 5-4. *Weather briefing shorthand.*

IPW	ice pellet showers
K	smoke
L	drizzle
R	rain
RW	rain showers
S	snow
SG	snow grains
SP	snow pellets
SW	snow showers
T	thunderstorms
ZR	freezing rain
ZL	freezing drizzle

Minus sign (−) indicates light, plus sign (+) indicates heavy.

ADVERSE CONDITIONS

The following briefing is quoted from an actual briefing at Westchester County Airport one November day not too long ago.

The briefer reads the adverse conditions from his computer screen, then proceeds in sequence through the other items on the briefing checklist:

"Covering your route of flight this morning we have SIGMET November Six for occasional severe icing in clouds and precipitation above ten thousand feet.

"We have AIRMET Oscar Four for occasional moderate rime icing from the freezing level to fourteen thousand feet and Oscar Seven for occasional moderate turbulence below ten thousand and moderate to severe low level winds. And here's a NOTAM—Binghamton radar is out of service.

"Westchester at zero niner hundred had indefinite one hundred feet obscured, one quarter mile in light rain and fog. Temperature four nine, dew point four six, winds calm.

"Along your route there are scattered to broken clouds below one thousand feet; then two to three thousand broken and four to five thousand broken, variable overcast.

"Binghamton at zero niner hundred was six hundred feet scattered, two thousand broken, visibility twelve miles. Temperature four three, dew point four one, winds three zero at nine knots.

"We have a PIREP from seven miles west of Kingston VOR from a B-E thirty-three at five thousand feet reporting a base of scattered clouds at one thousand two hundred feet, broken at twelve thousand, visibility five to ten miles. Temperature plus eight, occasional light turbulence."

FORECASTS

"Westchester prior to eleven hundred local is forecasting seven hundred overcast, two miles in fog. Wind three two zero at seven. Occasionally ceilings of one thousand one hundred feet overcast, four miles in light rain and fog. After eleven hundred local, you can expect two thousand overcast variable three thousand five hundred overcast, visibility better than six, wind three one zero at ten.

"En route, Poughkeepsie is forecasting the same improving conditions.

"Binghamton between ten and twelve hundred local is forecasting five thousand scattered, one thousand five hundred overcast, visibility more than six miles, wind two six zero at nine. Occasionally five hundred broken, two miles in fog, chance of light rain showers.

"After twelve hundred Binghamton is forecasting one thousand two hundred scattered variable broken, two thousand broken, visibility better than six miles, wind two seven zero at twelve knots. Chance of light rain showers."

WINDS ALOFT

"Winds at three, six, and niner thousand feet at Kennedy are: one eight zero at two four, one niner zero at two eight, and two zero zero at three two, with a temperature of plus three.

"At Wilkes-Barre at three, six, and niner thousand feet, the winds are two three zero at six, two zero zero at two seven, and two zero zero at four two, with a temperature of plus one.

"The freezing level at Kennedy is between ten and eleven thousand feet. At Binghamton the freezing level is nine to ten thousand feet."

ALTERNATE

Binghamton is not forecast to be above the "VFR plus 1,000" rule of thumb for requiring an alternate because the ceiling is too low; an alternate airport is needed.

"Wilkes-Barre looks good. They are currently reporting an estimated twenty thousand broken, twenty five miles. Temperature four nine, dew point four two, winds two eight zero at thirteen. After ten hundred, for the rest of the day, Allentown is forecasting four thousand five hundred broken, variable scattered, visibility better than six, winds two six zero at ten."

Not only is Wilkes-Barre a good alternate, but it's also a convenient, easily located airport. Now we have all the information needed to make a go/no-go decision, and we have jotted it all down on our weather briefing form (FIG. 5-5).

GO OR NO-GO?

What does all this mean? It is "go" or "no-go?"

Return to the checklist of five go/no-go factors and ask whether any will prevent flight as planned:

1. Thunderstorms.
2. Turbulence.
3. Icing.
4. Fog.
5. Departure and destination minimums.

Thunderstorms

No mention of thunderstorms in SIGMETs, AIRMETs, reports, or forecasts. "Go."

1. ADVERSE CONDITIONS
2. SYNOPSIS
3. CURRENT CONDITIONS
4. ENROUTE FORECASTS
5. DESTINATION FORECASTS
6. WINDS ALOFT
7. NOTAMS
8. ATC DELAYS

11/30 N3456X
IFR HPN - BGM
ETO 1100 local
+5
1600 UTC

1. Sig NG - occ svr icing in clds + precip above 10,000
   Airm Ø4 - occ mod rime ic frzng lvl to 14,000
   Airm Ø7 - occ mod trblnc blo 10,000   mod/svr
      low lvl wnds
   Notam - BGM radar OTS

3. HPN - Ø9ØØ W⊥X⊹ -RF 49/46 clm
   Enroute - sct v bkn blo 10, 20-30 bkn, 40-50 bkn v
      ovc
   BGM Ø9ØØ 6 sct 20 ovc 12 43/41 300/19
   Precip 7 E Kingston BE 33 @ 5000 Base 12 sctd
      120 sctd 10+   8c

4. HPN to 11ØØ - 7 ovc 2F 320/7 occ cigs 11 ovc 4 -RF
      aftr 1100 - 20 ovc 35 ovc 6+   310/10
   POU - same
   BGM - 10-12ØØ 5 sctd 15 ovc +6 260/9 occ 5 sctd/2F
      chnc RW
      12ØØ 12 sctd v bkn 20 bkn 6+ 270/12
      chnc RW

6. JFK 3 - 180/24 6 - 190/28 9 - 200/32+3
   Wilkes B 3 - 236/6 6 - 200/27 9 - 200/42 +1

   alt - AVP 200 bkn 25 49/42 280/13
      aftr 1000 45 bkn v sctd 6+ 260/10

Fig. 5-5. *Details of the weather for the IFR flight from Westchester County to Binghamton, as copied over the telephone.*

Turbulence

The briefer mentioned an AIRMET for occasional moderate turbulence below 10,000 feet, a PIREP for occasional light turbulence in the vicinity of Kingston, and winds aloft of 32 and 42 knots at 9,000 feet. AIRMET Oscar Seven is a concern with moderate to severe low level winds.

But turbulence reported in the AIRMET and the PIREP would not damage the airplane. The flight progresses into improving weather, so any turbulence-induced fatigue factor should not be as important as if we were heading into deteriorating weather.

Surface winds at Westchester and Binghamton should range from calm to 13 knots and that's manageable. "Go."

Icing

Icing is mentioned in SIGMET November 6 and AIRMET Oscar 4. A westbound (even thousand) cruise altitude of 8,000 feet will stay below the freezing level. "Go."

Fog

Westchester reported dense fog at 0900, but it seems to be dissipating. There is no fog at Binghamton nor at Wilkes-Barre, the alternate. Go, but delay takeoff at Westchester until 1100 so the ceiling and visibility will be sufficient for an instrument approach if an emergency return is necessary.

Minimums

Departure minimums will be a "go" after 1100. Binghamton is forecast to be well above minimums when we arrive. "Go."

Practice this reasoning prior to each instrument flight and it will become easier to weigh the major factors in the go/no-go decision with solid evidence for the decision. Occasionally the judgement calls will be too close for comfort, or the weather will unpredictably improve one hour and worsen the next hour. We have all seen days like that. For doubtful situations like these, apply the old pilot's rule of thumb: "If in doubt, wait it out."

In this case it's a Go. Complete the flight log by entering estimated ground speeds, computing the time en route for each leg, and supplying additional details.

ESTIMATED CLIMBOUT TIME

Sometimes instrument students are perplexed about how to compute an accurate time en route to the first fix because it involves takeoff, turns, and climbs. I teach students a simple and surprisingly accurate method to figure time to the first fix. Measure the distance from departure to the first fix, estimate winds from the forecast winds aloft reports, compute ground speed and time en route, and add one minute for each 2,000 feet of climb.

This will be very close to the actual time to reach the first fix. If an airplane is a slow climber, such as a Cessna 152, consider adding one minute for each 1,000 feet of climb.

Absolute precision on the first leg is not that important. ATC won't enforce the ± three minute standard for arrival times because they may amend the clearance with additional turns and changes in altitude. An accurate time en route to the first fix will ensure that the time en route for the entire flight will also be reasonably accurate.

WIND AND GROUND SPEED

En route to Binghamton, the distance to the first fix, HAAYS intersection, is a shade over 37 nautical miles. Kennedy—the weather observation point nearest our departure airport—was reporting these winds:

180/24 at 3,000 feet.
190/28 at 6,000 feet.
200/32 at 9,000 feet.

Select winds at 6,000 feet—190/28—as a good compromise for the climbout to 9,000 feet. Allow 90 knots for the climb speed on a course of 325° and the wind of 190/28 will yield a ground speed of 108 knots. This ground speed computes to 21 minutes en route to HAAYS. A minute for each 2,000 feet of climb adds 4.5 minutes to reach 9,000 feet. Time en route to HAAYS is 21 + 4.5 = 25.5 minutes. Round that off to the nearest even number and enter 26 minutes for the first leg. Use Wilkes-Barre winds at 9,000 feet (200/42) for the remainder of the flight to Binghamton. Ground speed for the approach should be 90 knots.

Calculator Options

Most pilots learned to figure wind problems on the E6B computer. This device is perfectly satisfactory for flight planning calculations. A number of aviation-oriented electronic calculators, which some pilots prefer to use, include the Avstar, Navigator, and Prostar.

Other pilots do most calculations mentally, using "Rules of Thumb" shown in FIG. 5-6. Decide on a system and stick to it.

Instructor Note. Understand all calculation methods—E6B, electronic calculators, and rules of thumb—because you never know which system a student will prefer to use.

Enter the Kennedy (JFK) and Wilkes-Barre (AVP) winds in the lower right corner of the flight log as shown in FIG. 5-7. While working in this area fill in the VFR WX AT: section for reference, if necessary.

Fig. 5-6. *"Rules of thumb" calculations for time to climb, ground speed, time to fly, and wind correction angle.*

TIME TO CLIMB

1. Estimate distance to reaching altitude. (Include vector, reversal, or circle.)
2. Use filed true airspeed for time/distance calculation.
3. Add one minute for each 2,000 feet of climb.

GROUND SPEED

Crosswind Angle	Effect on True Airspeed
0–15°	Full value of wind speed
30°	.9 wind speed
45°	.7 wind speed
60°	.5 wind speed
75°	.3 wind speed
90°	.1 wind speed

TIME TO FLY

Drop the last digit of the air speed (ground speed, if known); you will fly that many miles in 6 minutes.

Examples:

Speed 150 - 15 miles every 6 minutes.

Miles to go - 7. Time to fly - 3 minutes.

Miles to go - 20. Time to fly - 8 minutes.

Speed 90 - 9 miles every 6 minutes.

Miles to go - 5. Time to fly - 3 minutes.

Miles to go - 35. Time to fly - 24 minutes.

WIND CORRECTION ANGLE

True Airspeed	Crosswind Angle	Wind Corr. Angle
100 - 120 - 140	90°	½ wind speed
	45°	⅓ wind speed
150 - 180 - 200	90°	⅓ wind speed
	45°	¼ wind speed

Compute and fill in the estimated ground speeds and times en route as in FIG. 5-8. The estimated total time en route adds up to one hour and eight minutes, which is written as 1 + 08.

FLIGHT PLAN TO ALTERNATE

Plan a diversion to the alternate, Wilkes-Barre. Weather may be improving, but always be prepared for diversion to an alternate airport. (If you forget to carry an umbrella on a threatening day, it is sure to rain.)

				TWR	117.3
				GRND	121.9

WINDS ALOFT

Sta.	A°	Dir.	Vel	Temp	A	Dr	VI	T	A	D	V	T
JFK	3	18	24	-	6	19	28	+8	9	20	32	+1
AVP		23	6	-		20	27	+7		20	42	+1

VFR WX AT: **AvP (Wilkes-Barre)**
0900 - 200 ⊕ 25 49/42 280/13

TANK ON	TANK OFF	TANK ON	TANK OFF
1505			

Fig. 5-7. (above) *Filled-in winds aloft and nearest VFR weather blocks on flight log form for IFR flight to Binghamton.*

Fig. 5-8. *Ground speed and estimated time calculations for IFR flight to Binghamton.*

DIST.	G/S	TIME ENR	TIME ARR
TO	EST	EST	EST
REM	ACT	ACT	ACT
37+	108	26	31
		28	33
8	144	4	37
		5	38
66	134	30	08
		28	06
14	109	8	14
		9	15
125		68	
		(1+08)	

57

Plan for V149 to Lake Henry VOR (LHY), which is also the initial approach fix (IAF) for the Wilkes-Barre approach that will most likely be in use, ILS 22. Radar vectors to the ILS 22 final approach course are likely, but plan to execute the complete approach in case radio communication is lost or the approach control radar fails.

Identical methods for computing the times for climbout, en route, and descent, reveal results shown in FIG. 5-9. Add the times en route from Westchester to Binghamton and thence to Wilkes-Barre plus the 45 minutes required by regulation (FAR 91.23) to obtain a total time for the flight, including alternate (plus 45 minutes), of 2 hours and 32 minutes, well within the fuel range of the aircraft. The completed flight log is shown in FIG. 5-10.

IN-FLIGHT NOTATIONS

Note the ample amount of space remaining for entering in-flight items. Locate where the following in-flight information goes:

- Actual ground speed and actual time en route.
- Estimated and actual times of arrival (ETA and ATA).
- Position report sequence (lower left) if needed.
- Space for clearances (center) and below that more space for clearance limits.
- Takeoff and landing runway information, and which approach might be expected.
- Time off, time on, total time en route, and tachometer (or Hobbs) reading at the beginning and end of the flight.
- Fuel management logs (lower right).

FILING THE FLIGHT PLAN

The completed flight plan is shown in FIG. 5-11.

Use the equipment code in FIG. 5-12; add the code to the briefing card for later reference if a piece of equipment fails, or you are flying a different airplane.

Student Note: Use the instructor's name, address, and telephone number because a student cannot legally file an IFR flight plan. The exception is the instrument flight test.

When filing by telephone stay on the line for a specialist to copy the flight plan rather than simply recording your message. Request information that has been updated since the previous briefing.

The FAA recently stated in *FAA Aviation News*: "Live briefers are a good source of unpublished Notices to Airmen concerning important data about airport of runway closures, military flight route training activity, obstructions to flight, outages and shutdowns, etc. The latest pilot reports of weather conditions aloft (PIREPs) are also more likely to be available from FSS specialists than from pre-recorded messages."

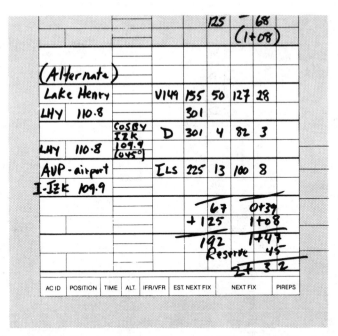

Fig. 5-9. *Planning for alternate, Wilkes-Barre, on IFR flight to Binghamton.*

ABBREVIATED BRIEFINGS

Certain "magic words" that I will introduce from time to time simplify instrument procedures; one very handy pair is "abbreviated briefing."

Request an abbreviated briefing for:

- Updated information, such as NOTAMS, to supplement TWEB or PATWAS recordings.
- Updated specifics of a previous full-length briefing.
- One or two specific items. This would be the case, for example, when filing IFR on a "severe clear" day to a nearby airport that you are familiar with. Request a terminal forecast to make sure VFR conditions will continue through your ETA.

Try abbreviated briefings a couple of times during training when weather conditions are favorable. This procedure can save a lot of time when used properly.

OUTLOOK BRIEFINGS

Another good pair of magic words is outlook briefing. Request this service for a short, live briefing to supplement other sources regarding a departure time more than six hours away.

DATE November 30   PILOT: Sollman/Harris

FLIGHT PLAN ROUTE

Westchester 9 Haays △
D-Huo V252 Lathy △

ATC CLEARS N 3456 X

C - BGM W-9 Haays F
Ex 99/10 120.55 1761

TAS: 125	IAS:		ASGND ALT
RPM: 2300	MP: 23'		

	CALL SIGN	FREQ.	ASGND ALT
	ATIS	133.8	
TAKEOFF: RWY WIND KTS ALT	CPT	129.25	30
	GRND	121.8	90
Field Elev: 439	TWR	119.7	80
ALTIMETER ERROR ☐	DC	120.55	40
LANDING: Field Elev: 1630 RWY WIND KTS ALT	BOS	120.2	
	NY	128.5	
TYPE OF APPROACH:			
TIME ON:	ATIS	128.15	
TIME OFF:	AC	118.0	
TOTAL TIME ENROUTE:	TWR	114.3	
TACH ON TACH OFF	GRND	121.9	

WINDS ALOFT

Sta.	A Dir.	Vel	A Temp	A Dr	A T	V	A D	V	T
JFK	3 18	24	-	6 19	28 +8	9	20 52	+1
AVP	23	6	-	20 27+7	20 42	+1

VFR WX AT: AVP (Wilkes-Barre)
0900 - 200 ① 25 49/42 2⑩/13

	TANK ON	TANK OFF
TANK OFF	TANK ON	

FIX	INTERSEC	ROUTE	M.C.	DIST.	G/S	TIME ENR	TIME ARR		
NAME	NAME IDENT	VIA	TO FROM	TO REM	EST ACT	EST ACT	EST ACT	CLEARANCE LIMIT	
IDENT. FREQ	FREQ RADIAL	ALT							

—	Haays STW 106.6 054	W-9	305 325	37+	108	26		
Huo 116.1						28		
Huguenot	V273	325	8	144	4			
Huo 116.1	314				5			
BGM 112.2	Lathy ANY 116.9 232	V252	318	66	134	30		
						28		
BGM airport	ILS	340	19	149	8			
I-BGM 110.3						9		
			125			68		
					(1+08)			
(Alternate)								
Lake Henry		V149	155	50	127	28		
LHY 110.8			301					
	Cosby IZ6 109.9 045-0	D	301	4	82	3		
LHY 110.8		ILS	225	13	100	8		
AVP - airport				67	0+39			
I-IZ6 109.9			+125		1+08			
			192		1+47			
			Reserve		45			
					24			
					3 12			

AC ID	POSITION	TIME	ALT	IFR/VFR	EST NEXT FIX	NEXT FIX	PIREPS

Fig. 5-10. *The complete log for IFR flight to Binghamton showing all details that can be filled in prior to takeoff.*

60

Form Approved: OMB No. 2120-0026

U.S. DEPARTMENT OF TRANSPORTATION FEDERAL AVIATION ADMINISTRATION **FLIGHT PLAN**	(FAA USE ONLY) ☐ PILOT BRIEFING ☐ VNR ☐ STOPOVER		TIME STARTED	SPECIALIST INITIALS

1. TYPE	2. AIRCRAFT IDENTIFICATION	3. AIRCRAFT TYPE/ SPECIAL EQUIPMENT	4. TRUE AIRSPEED	5. DEPARTURE POINT	6. DEPARTURE TIME		7. CRUISING ALTITUDE
☑ VFR ☐ IFR ☐ DVFR	N3456X	C R2/A	125 KTS	Westchester County	PROPOSED (Z) 1600Z	ACTUAL (Z)	8,000

8. ROUTE OF FLIGHT

Westchester 9 Haays △
D HUO V252 LaHy △

9. DESTINATION (Name of airport and city)	10. EST. TIME ENROUTE		11. REMARKS
	HOURS	MINUTES	
BGM	1	08	

12. FUEL ON BOARD		13. ALTERNATE AIRPORT(S)	14. PILOT'S NAME, ADDRESS & TELEPHONE NUMBER & AIRCRAFT HOME BASE	15. NUMBER ABOARD
HOURS	MINUTES	Wilkes- Barre	(Instructor's name, address, etc.)	2
5	30		17. DESTINATION CONTACT/TELEPHONE (OPTIONAL)	

16. COLOR OF AIRCRAFT	CIVIL AIRCRAFT PILOTS. FAR Part 91 requires you file an IFR flight plan to operate under instrument flight rules in controlled airspace. Failure to file could result in a civil penalty not to exceed $1,000 for each violation (Section 901 of the Federal Aviation Act of 1958, as amended). Filing of a VFR flight plan is recommended as a good operating practice. See also Part 99 for requirements concerning DVFR flight plans.
White, blue trim	

FAA Form 7233-1 (8-82)   CLOSE VFR FLIGHT PLAN WITH_____ FSS ON ARRIVAL

Fig. 5-11. *The completed flight plan for the IFR cross-country to Binghamton.*

/X	No transponder.
/T	Transponder with no altitude encoding capability.
/U	Transponder with altitude encoding capability.
/D	DME, but no transponder.
/B	DME and transponder, but no altitude encoding capability.
/A	DME and transponder with altitude encoding capability.
/M	TACAN only, but no transponder.
/N	TACAN only and transponder, but with no altitude encoding capability.
/P	TACAN only and transponder with altitude encoding capability.
/C	RNAV and transponder, but with no altitude encoding capability.
/R	RNAV and transponder with altitude encoding capability.
/W	RNAV but no transponder.

Fig. 5-12. *ATC equipment codes for use on flight plan.*

Outlook briefings are particularly helpful when it's harder than usual to outguess the weather. There is no way to know when a stalled low pressure system may start moving unless a pilot has access to winds aloft charts, constant pressure charts, and other highly technical information. Even then, the charts may not be much help.

THE ONE-CALL TECHNIQUE

Another method to speed up the weather briefing and filing process is combining the weather briefing and flight plan filing in one telephone call.

Completion of phase one flight planning—elements completed prior to calling flight service—will reveal all details to complete the flight plan form, except estimated time en route and alternate airport.

The weather briefing will quickly establish whether or not an airport is necessary and the briefer will assist with the selection.

Estimated time en route to Binghamton is based upon a true airspeed (TAS) of 125 knots for the Cessna 182. It's a simple matter to bracket this airspeed in increments of 10 knots on either side and make a series of time-speed-distance computations.

Quick Estimates

Before calling flight service compute several figures for time en route using 125 total nautical miles en route. Bracket the 125 knot TAS with headwinds of 30, 20, and 10 knots and tailwinds of 10, 20 and 30 knots to reveal:

$$
\begin{array}{llll}
125 & @ & \ 95 & = \ 1 + 19 \ (30 \text{ knot headwind}) \\
125 & @ & 105 & = \ 1 + 12 \ (20 \text{ knot headwind}) \\
125 & @ & 115 & = \ 1 + 05 \ (10 \text{ knot headwind}) \\
125 & @ & 125 & = \ 1 + 00 \ (\text{zero wind factor}) \\
125 & @ & 135 & = \ 0 + 56 \ (10 \text{ knot tailwind}) \\
125 & @ & 145 & = \ 0 + 52 \ (20 \text{ knot tailwind}) \\
125 & @ & 155 & = \ 0 + 38 \ (30 \text{ knot tailwind})
\end{array}
$$

When the briefer gives the winds, simply pick the closest estimate of ground speed. A tailwind off the left quarter during this flight probably means a tailwind component of approximately 20 knots, based on the forecast winds for Wilkes-Barre. This produces a time en route of 52 minutes.

Add one minute for every 2,000 of climb to our planned altitude of 8,000 feet and you arrive at a total time en route of 56 minutes. (You can simplify the climb estimate by automatically adding five minutes for climbs up to 10,000 feet and 10 minutes for climbs over 10,000 feet.)

TOTAL TIME EN ROUTE

Return to the flight log and see the estimated 68 minutes for total time en route. Is the 56 minute quick estimate close enough for filing purposes? Certainly. Time en route establishes when ATC expects the approach to begin in the event of two-way communications failure.

Emergency procedures will be covered in greater detail later, but let's briefly consider lost communications. If two-way radio communication is lost, ATC will expect a pilot to carry out the lost communications procedure in FAR 91.127, "IFR Operations; Two-Way Radio Communications Failure."

Fly to the destination following the last assigned routing and altitude. "Begin descent from the en route altitude or flight level upon reaching the fix from which the approach begins [Initial Approach Fix], but not before—

"The expect-approach-clearance time (if received)"; or "If no expect-approach-clearance time has been received, at the estimated time of arrival, shown on the flight plan, as amended with ATC." ATC will expect an arrival time as filed. Enter a holding pattern if early. If late, ATC will protect the approach airspace for 30 minutes, then begin lost aircraft procedures.

The quick estimate indicates early arrival at Binghamton, which is perfectly all right. The approach would begin out of the holding pattern when the time en route expired.

If the estimated time en route changes either way by more than five minutes, notify ATC about a revised ETE or ETA. Not only is this good insurance to cover the possibility of two-way radio communication failure, but it will also help ATC efficiency.

Don't use the one-call procedure if concerned about any of the five go/no-go weather factors. Get a thorough briefing and study the impact of weather on the flight, then make a second call to update the weather information and file the flight plan.

THE FAST FILE PROCEDURE

Certain times it is perfectly all right to brief yourself and file a recorded flight plan without talking to a specialist. This is known as "fast file."

When planning a training flight, for example, and the intention is to cancel IFR at the practice area, call fast file and record the flight plan without talking to a briefer. Don't rely on a recorded process if any of the five go/no-go weather factors is questionable. Apply the old pilots' rule of thumb: "When in doubt, check it out!"

Recorded Flight Plan Tips

Several simple steps can ensure that a recorded flight plan is processed quickly and a clearance is ready at departure.

- Use "Fast File One" for IFR (and VFR) flight plans when the departure time is one hour or less from the current time and when cancelling or amending previously filed flight plans.

- Use "Fast File Two" for flight plans when the departure time is one hour or more from the current time.
- Speak slowly and enunciate clearly. Read some flight plans into a tape recorder and play them back to hear yourself.
- Seven minutes are available to read the flight plan. It should be more than enough time. Make sure all the details of the flight plan form are completed and ready to read.
- Mention that the pilot (or instructor) and airplane are IFR rated, and give the source of the weather briefing: "Full weather briefing by 1-800-WX-BRIEF specialist" or "Weather briefing by PATWAS," and so forth. Provide these two items because FSS might delay entering the flight plan into the computer until a telephone call verifies that legality of the IFR flight.
- Always provide a telephone number—with area code—where you can be reached until departure time. One of the most common causes of "lost" flight plans, says the FAA, is missing or incorrect information. If flight service can't get back to a pilot when a problem arises with a flight plan, they will be unable to process it.

ATC has magic words too and, if you have done a thorough job of flight planning, you are more likely to hear the welcomed words "Cleared as Filed" when the clearance is read to you.

Phase one planning covered all details prior to contacting flight service—route, destination, departures, approaches, frequencies, and the big picture of the weather.

Phase Two was either a two-call, a one-call, or a recorded communication with flight service to obtain the detailed weather necessary to make a go/no-go decision and to file the flight plan.

Follow this approach to planning an IFR flight from the very first day and the planning becomes easier, faster, and much more interesting. Following this procedure and filing an IFR flight plan for every instrument training flight—even if it's just out into the local area for practice in basic maneuvers—will provide a better chance of passing the instrument check ride.

Save all logs and flight plan forms. It was hard work to get all the information necessary for those flights, so save the information and use it again. Always check the information against the current charts for changes. Use routes and approaches that worked well on previous flights and especially remember any amended clearances received. Amended clearances might suggest better routes or fixes that can be incorporated into future flight planning.

6
Airplane, Instrument, and Equipment Checks

Y OU HAVE COMPLETED THE FLIGHT LOG, CHECKED THE WEATHER AND FOUND THAT the flight can be safely completed. The flight plan is filed and the pilot is prepared but what about the airplane? Is it ready for an IFR flight in actual instrument weather?

Several items on the preflight checklist require special attention for an IFR flight; see FIG. 6-1 for a basic list. Consider adding items for a specific airplane and equipment. (Note: Some of the items will be discussed in detail in the next chapter.)

FUEL QUANTITY

Some checks are obvious. Turn on the electrical power and check fuel gauges to make sure there is enough fuel to fly to the destination plus 45 minutes at normal cruising speed. If an alternate airport is required, verify enough fuel to the destination, then to the alternate plus 45 minutes. This item is first on the checklist so the pilot can continue the checklist while waiting to refuel.

It's always a good practice to start an instrument flight with full fuel tanks if weight permits, visually confirmed. Nothing is as useless as the ''runway behind, the air above, and the fuel in the fuel truck!''

During a round-robin at each stop ask, ''Should I top off again?'' If in doubt, top off. There is no reason to worry about running out of fuel in addition to all the other concerns.

ATIS

While the electrical power is on, tune in the ATIS frequency if available at the departure airport, and write down the current information. Set the altimeter to the current altimeter

Fig. 6-1. *IFR checklist items.*

PRE-TAXI CHECK

Before starting:
- ☐ Outside antennas - all secure
- ☐ Fuel - to destination, alternate, plus 45 minutes
- ☐ ATIS - copied
- ☐ Altimeter - set, error noted
- ☐ Airspeed, VSI - both on zero
- ☐ Magnetic compass - shows correct heading
- ☐ Clock - running and set correctly
- ☐ Pitot heat - working
- ☐ Lights - all working
- ☐ VOR - checked within 30 days
- ☐ Charts and logs - sequenced

After starting:
- ☐ Ammeter - checked
- ☐ Suction gauge - normal
- ☐ COM radios - all checked and set in sequence
- ☐ NAV radios - all checked and set for departure
- ☐ Marker beacon lights - checked
- ☐ Heading indicator - set
- ☐ Attitude Indicator - normal
- ☐ Alternate static source (if any) - working
- ☐ Clearance - copied
- ☐ Transponder - code set, checked, turned to standby

TAXI CHECK

- ☐ Heading indicator - responding normally to turns
- ☐ Attitude indicator - normal and stable
- ☐ Turn coordinator - responding normally to turns

PRE-TAKEOFF CHECK

- ☐ Approach charts - emergency return chart on top
- ☐ NAV radios - double check set for departure
- ☐ COM radios - Departure control on #2
- ☐ Heading indicator - aligned with runway centerline

RUNWAY ITEMS - "STP"

- ☐ Strobes (or rotating beacon) - on
- ☐ Transponder - on
- ☐ Pitot heat - on

---

setting as reported by ATIS. If there is a difference of more than 75 feet between what the altimeter reads and the ramp elevation where the plane is parked, don't go. There is something seriously wrong with the altimeter.

Reference the altimeter with the ramp elevation where the plane is parked, not the general field elevation. Airports are seldom level. Look at the airport diagram on the approach chart and select the elevation nearest the parking ramp. As noted in the flight planning chapter, there can be considerable difference between official airport elevation and ramp elevation.

ALTIMETER ERROR

There is frequently confusion about what action to take with altimeter error. Use the given barometric pressure and note the error on the flight log in the appropriate spot. Fly the altitudes shown on the altimeter and disregard the error until the approach.

Add the error to published minimums, regardless of whether this error is plus or minus. For example, the White Plains Decision Height (DH) on the ILS 16 approach is 639 feet. If the altimeter error is 50 feet, add it to the minimum and use a DH of 689 feet regardless of whether the error is plus 50 feet or minus 50 feet. Always add the difference to be on the safe side.

Check the airspeed indicator and the vertical speed indicator, which should indicate zero and zero. Most vertical speed indicators have an adjustment screw on the lower left corner of the instrument for calibration to zero.

If there is no adjustment screw, allow for the error when interpreting indications in flight. For example, if the instrument shows a 50 foot-per-minute climb on the ramp, that is level flight in the air.

ELECTRICAL EQUIPMENT

Turn on and check the pitot heat and all outside lights, including landing and taxi lights, even if the flight will occur during daylight hours. Pitot heat is necessary if any part of the flight is in the clouds, especially when climbing through a cloud layer after takeoff. It's a very strange feeling to see airspeed drop toward stalling speed during an instrument climbout because ice is building up in the pitot tube!

Lights might be necessary for visual identification of the airplane by tower controllers during an approach in low visibility.

VOR CHECKS

Verify that the record of VOR checks is current. The VOR check is probably the most overlooked or ignored regulation. There is nothing mysterious about the check, nor is the regulation hard to understand. FAR 91.25 says that no person may use the VORs on an IFR flight unless they have been checked within the preceding 30 days and found to be within certain limits.

A record of VOR checks must be kept in the plane. If no log entry is found attesting to a VOR check within the preceding 30 days, it is illegal to make an instrument flight unless the VORs are checked before takeoff. In spite of the legal requirements, it doesn't make sense to depart on an instrument flight without knowing that VORs are accurate and within limits.

If the VORs have been checked within the 30-day limit everything is OK. If not, determine if a VOR check is possible before taking off. VOR checks will be explained later. The point here is that checking the status of the VORs is an important item on the IFR preflight inspection.

COCKPIT ORGANIZATION

The basic principle of cockpit organization is two-fold: (1) organize charts and flight log to avoid fumbling for a vital piece of information, and (2) check and set up both communications and navigation radios ahead of time according to the expected sequence of frequencies.

One very real hazard associated with cockpit confusion is vertigo. Sudden head movement while searching for a chart might cause an attack of vertigo. It might be more severe when reaching back to pick up something from a flight case on the rear floor, then snapping to an upright position.

How to Sequence Logs and Charts

My favorite device for holding charts and logs is a standard 8½″ × 11″ clipboard or folder. A folding flight desk that rests on the lap, or a knee board that straps to a thigh works well.

I place the flight log and flight plan on top. Underneath those I place the en route chart folded so I can see the route and airways or fixes that might be in any amended clearance.

Approach charts for the departure airport are placed under the en route chart in case I have to return for an instrument approach shortly after takeoff. The loose pages are reinserted in the NOS booklet later. Photocopy commonly used approach charts and label the copy with the expiration date.

Organizing Approach Charts

Next, tear out all approach charts for the destination airport and place them underneath the charts for the departure airport. Put the expected approach on top. Finally, tear out and insert all approaches for the alternate airport.

RADIO PREPARATION

Before starting an engine, determine what sequence of frequencies will be used for communications and navigation. Write the frequencies in sequence, starting from the top. A top-to-bottom and back sequence used consistently will reduce the chances of selecting the wrong radio at the wrong time.

In most cases the departure communications sequence will be:

- ATIS
- Clearance delivery or pre-taxi clearance
- Ground control
- Tower
- Departure control
- Air route traffic control center

Tune the ATIS frequency first on the number one communications radio. Move down and tune clearance delivery. (Clearance delivery is often handled by ground control. Some very large airports might also have a gate control frequency. Airline pilots monitor the gate control frequency for clearance to start their engines and save fuel.)

Listen to transmissions on the first frequency, adjust the volume to a comfortable level, and turn the squelch, or sensitivity until the "noise" begins. Then back off on the squelch until the "noise" disappears.

Follow the same procedure with the second radio in the communications sequence.

Presetting Frequencies

Some newer NAV/COMs will store up to nine additional frequencies. I once preset NAV frequencies to fly all the way from Palm Beach, Florida, to Myrtle Beach, South Carolina, without dialing in a new VOR frequency. The same features are also available on the COM side. I frequently pre-program the ATIS, approach, tower, and ground control at my destination airport. It surely cuts down on the workload in the air.

Use the top-to-bottom approach to check navigation radios as well. I check the VOR receivers' accuracy on every flight. Look in the "VOR Receiver Check" section in the back of the Airport/Facility Directory (A/FD) and see which checks are available at your home airport as well as other airports frequently used (FIG. 6-2). Set up the test frequencies before starting, like the communications radios.

VOR Checks with VOT

The best method of checking VORs is with a VOR Test Facility (VOT). The VOT is a special VOR ground facility that transmits only the 180 degree radial. Tune in the VOT frequency listed in the A/FD and dial in 180 with the omni bearing selector (OBS) knob. The needle should center in the TO position. An easy way to remember this is "Cessna 182." The C in Cessna is for center, one-eighty gives you the radial, and two is the direction. The maximum permissible bearing error is ± four degrees with the VOT.

Unfortunately, only a limited number of airports in the United States have a VOT. Many are very busy, high traffic hubs—such as Kennedy, LaGuardia, and Los Angeles— where instrument training flights are not always practical. If you don't have access to a VOT, the next best check is with a VOR located on the field or close enough to receive

VOR RECEIVER CHECK
WEST VIRGINIA

VOR RECEIVER CHECK POINTS

Facility Name (Arpt Name)	Freq/Ident	Type Check Pt. Gnd. AB/ALT	Azimuth from Fac. Mag	Dist. from Fac. N.M.	Check Point Description
Clarksburg (Benedum)	112.6/CKB	G	039		Taxiway charlie in front of terminal building.
Martinsburg (Eastern West Virginia Regional Shepherd Fld)	112.1/MRB	G	281		At apch end Rwy 35.
		G	285	6.3	Near intersection Rwy 26 and taxiway B.
		G	287		Near hold line for Rwy 26.
Morgantown (Morgantown Muni-Walter L. Bill Hart Fld)	111.6/MGW	G	335	6	On N end of terminal ramp.
Parkersburg (Wood County Gill Robb Wilson Fld)	108.6/JPU	G	207		On runup area N side of Rwy 28.
	108.6/JPU	G	211		On corner of twy to Rwy 21.
Wheeling (Wheeling Ohio Co)	112.2/HLG	G	223		On taxiway on east side of ramp.
White Sulphur Spgs (Greenbrier)	108.4/SSU	G	306	1.7	On ramp 100' N of Rwy 26 centerline in front of hangar.

VOR TEST FACILITIES (VOT)

Facility Name (Airport Name)	Freq.	Type VOT Facility	Remarks
Charleston Kanawha (Yeager)	108.8	G	

Fig. 6-2. *VOR receiver check information as published in* Airport/Facilities Directory.

on the ground. Tune both VORs to this station and note the indicated bearings to that station. The maximum permissible difference between the two bearings is four degrees.

Ground and Airborne VOR Checks

The A/FD also lists other ground and airborne checkpoints. Again, see what checkpoints are available. One disadvantage to ground checkpoints is that they may be located on the opposite side of the airport. And because airborne checkpoints require a cross-check between a VOR radial and some point on the ground, such as the end of a runway, they won't be much good on an IFR flight in the clouds!

Another method of checking VOR receivers is by tuning the same VOR in flight and noting the bearings from the station. The maximum permissible variation between the two bearings is four degrees. FAR 91.25, which covers VOR checks, also prescribes an airborne procedure for a single VOR receiver and for dual VOR receiver checks.

To be legal in meeting the 30-day requirement, VOR receiver checks must be logged with date, place, bearing error, and signature of the person making the check.

Take advantage of all opportunities to make airborne checks. Log them properly and keep the log in the airplane to avoid getting stuck on the ground on a routine IFR day simply because the VOR receivers have not had a legal check within the last 30 days.

ILS CHECK

If there is an ILS system at the departure airport, check out the ILS receiver after the VOR receiver check. Tune in the localizer frequency on the number one navigation radio, listen to the identification signal, and observe the needles for correct movement. The needles won't steady up in one position because the airplane is not aligned with the final approach course while at the parking ramp. If you hear a clear identification signal, the red warning flag on the instrument face disappears, and the needles are "alive," the receiver is operating properly.

Check the ILS identifier, then turn it off or turn it down to the lowest volume level and leave the localizer frequency on the number one receiver. You will be ready to make an emergency return shortly after takeoff without fumbling around for the correct frequency. Likewise, set the number two navigation receiver to the first en route VOR station and it is ready to proceed on course without cockpit confusion.

ADF CHECK

Get a good check of the ADF by tuning the frequency of the outer compass locator at the departure airport and observe the swing of the needle. When it steadies it should point toward the locator. With a little trial and error you can soon establish the relative bearing from the ramp to the locator and use this to check the accuracy of the ADF. Use the same procedure if the departure airport has an NDB approach and no ILS.

If there is neither a compass locator nor an NDB near the airport, use a local AM radio station to check the ADF. Try to find a station near the airport that is shown on a sectional chart and plot the bearing to the station and see how well the ADF needle matches that bearing.

TRANSPONDER CHECK

Finally, turn the transponder to the test position. If the indicator light comes on and blinks, it means that the transponder has run an internal check of its circuitry and everything is functioning normally.

However, be careful not to switch the transponder to the ON or ALT (Mode C) positions at this point. The reason is that in this great world of automation, the transponder signal may be sensed by the traffic control radar antenna and fed into the computer to indicate that an IFR flight has commenced. Leave the transponder in the STANDBY position until cleared onto the runway for takeoff and include the transponder in the final list of runway items.

TIPS TO REDUCE COCKPIT CONFUSION

The best ways to reduce cockpit confusion are:

- Organize logs and charts on a clipboard with the flight log on top and the other material in the proper sequence. Do this in the planning room before entering airplane.
- Tune the communications and navigation radio frequencies to check before starting the engine. Then use a top-to-bottom sequence for checking out each piece of radio navigation equipment. If frequencies checked are different from departure frequencies, tune in the departure frequencies, with the ILS of the departure airport, if available, on the number one navigation receiver. Then you can keep one step ahead throughout the flight.

All the checks above may be performed on battery power unless the temperature is below freezing. When it gets that cold, of course the battery has less power. It is a good idea to start the engine first, while the battery is still fresh, then perform the radio checks.

GYRO INSTRUMENTS

Other instrument checks can only be made after starting the engine. Prior to taxiing, set the heading indicator to match the heading on the magnetic compass. Next, check the attitude indicator; to ensure that it confirms level flight. (Pitch attitude cannot be set on the ground because there is no way of knowing when the airplane is sitting in a level attitude.)

Any drift by either instrument during taxi indicates a malfunction that should be diagnosed and repaired prior to IFR flight.

7
Clearances and Communications

COPYING CLEARANCES SEEMS TO BE A GREAT STUMBLING BLOCK IN THE MIND OF MANY instrument pilots. There is no need for this to happen. If you start out copying clearances the right way, you will soon find that this is one of the easier elements of instrument flight.

Because an IFR flight plan is filed for every training flight, you will have to copy IFR clearances from ATC—and read them back correctly—beginning with the first training flight. Based upon experience with hundreds of students, those with the best success in mastering clearance copying are those who are prepared to copy and read back clearances on every flight.

PRACTICE CLEARANCES

These tips will soon have you copying clearances like an airline pilot. But practice, practice, practice is the best way to become competent in copying clearances, especially in the beginning when the jargon is unfamiliar and the controllers seem to set new records for fast talk!

Once again I urge you to buy a multi-band portable radio with aviation bands or a transceiver to practice copying clearances delivered to other flights. The transceiver is a better investment because it can also be used to communicate with ATC if communications are lost while on an IFR flight plan. Other handy uses for a transceiver will be discussed later.

Read those clearances into a tape recorder then play the tape back and attempt to copy the clearances. It will be fun and it will develop a competence—within three or four lessons—that is unbelievable.

If you are unable to listen to actual clearances being delivered, ask an instructor to read simple clearances for practice.

Instructor Note. Why not make up an audio tape that students can use? A cassette recorder and a few short sessions in a parked aircraft on a busy day will produce a tape recording for students to use at home.

CLEARANCE SHORTHAND

It's impossible to copy a clearance in long hand and get it right. Every pilot develops a clearance "shorthand" that works well for him. Look over the list of simplified clearance shorthand symbols in FIG. 7-1 and practice using them. If a personal clearance shorthand is more comfortable and more in keeping with personal taste, be my guest and use symbols from the list that are helpful. All that's necessary is an ability to read back the clearance promptly and accurately.

On the first few instrument training flights tell ATC that you are an instrument student and ask the controller to "please" read the clearance slowly. When controllers realize that they are working with a student, they will read the clearance slower rather than read it over three or four times.

Be frank about expertise; ATC will cooperate in most cases. Controllers get in the habit of talking fast and they assume that any pilot can keep up with them. An FSS specialist will quickly request slower speaking if a pilot talks too fast when filing a flight plan.

This advice is not just for students because when an experienced pilot is at an unfamiliar airport expecting an unfamiliar clearance, he should request a slower delivery. ATC will usually cooperate.

Another tip on clearance copying: If the controller reads the clearance a little too fast and parts are missed, keep on writing and leave a blank spot for the missed information. Then read back what you have and ask for a repeat of any section that wasn't clear. Don't give up—keep on writing! The last thing clearance delivery wants to hear is: "Please repeat everything after 'ATC clears . . .'."

HANDLING AMENDED CLEARANCES

Now let's look at what's involved in copying clearances in the air. It's almost impossible to fly IFR anywhere these days without receiving an amended clearance or two—and they always seem to come as a surprise. Be prepared for them and avoid surprises.

Always have a blank spot on the flight log or an extra blank piece of paper handy to reach instantly when the controller issues a clearance.

Take a pencil and start writing on that piece of paper on the left hand edge while looking at the attitude indicator because it is the basic reference for control of the airplane. Keep the airplane straight and level.

Learn to copy clearances without looking at the paper by practicing on the ground. Try this experiment: Place a piece of paper on the clipboard, stare off into the distance

60	altitude - 6,000'
A	airport
<	after passing
>	before reaching
C	ACT clears
X	cross
D	direct
EAC	expect approach clearance
EFC	expect further clearance
F	as filed
H	hold
H-W	hold west, etc.
M	maintain (altitude)
O	VOR or VORTAC
RL	report leaving
RP	report passing
RR	report reaching
RV	radar vectors
RY Hdg	runway heading
LT	left turn after takeoff
RT	right turn after takeoff
V	Victor airways
til	until further advised
↑	climb to
↓	descend to
⊸	intercept
↕	shuttle climb
↕	shuttle descent

Fig. 7-1. *Clearance copying shorthand.*

at an object imagining it to be an attitude indicator, and write "I am a very good IFR pilot" several times, beginning from the left edge. It doesn't have to be perfect, and it's personal writing so you'll be able to read it, even if it runs uphill or downhill a little. Try this a couple of times to see how easy it is.

Now go back and play some of those recorded clearances. Copy them also without looking. Airborne clearances seldom cover more than two or three items at a time. Develop a clearance shorthand and practice copying full clearances on the ground without looking and there should be no difficulty copying the shorter airborne clearances.

For example, let's say that an amended clearance while airborne said: "Cessna three four five six Xray cleared direct Haays intersection maintain eight, report passing six." This is all you would have to write, using the clearance shorthand in FIG. 7-1: "D Haays M 80 RP 60".

OBTAINING CLEARANCES

After all this practice it is time to get on with the flight. An ATC clearance must be obtained.

Complete the preflight and get the cockpit organized before worrying about the clearance, then decide whether or not to start the engine before calling for a clearance. On a cold day it might be wise to start up, then call clearance delivery; engine and instruments can warm up while waiting for the clearance.

On the other hand, the delay might be extensive and the engine might run for a long time prior to taxi. If renting the plane according to the time on the Hobbs meter, this is a needless expense. Instead, call for the clearance then wait for it without the engine running. Today's transistorized radios use very little power and you can listen for up to 15 minutes or so and still have plenty of starting power, unless the battery is weak or the temperature is very cold.

Avoid this dilemma entirely by using a hand held transceiver to call clearance delivery. Once again, the battery powered transceiver is a backup radio in the event of a lost communications emergency.

When an IFR flight plan is filed, the information goes into an ATC computer. If the computer determines that the flight will not conflict with other traffic, it might be: "Cleared as filed."

If there is a conflict, ATC will resolve the conflict by assigning a different altitude, a different route, or by clearance to a fix short of the destination. If cleared short of the destination, you will receive an amended clearance to the destination when the traffic conflict has been resolved.

The flight plan normally remains in the computer until two hours after the proposed departure time. If the clearance is not requested by that time, the flight plan will be erased from the computer's memory.

Clearance "On Request"

Here is a point that many people misunderstand. ATC will not ordinarily get your clearance from the computer until requested. I have seen quite a few students sit on the ramp with their engine running waiting in vain for some message from ATC. It is the IFR pilot's responsibility to inform clearance delivery or ground control that he is ready to copy his clearance.

Pilot: "Cessna three four five six Xray IFR Binghamton with information Romeo" (the ATIS).

ATC: "Cessna three four five six Xray. Clearance on request."

"Clearance on request" means that the clearance delivery controller does not have your clearance in front of him, or that there is a problem with it that he may have to clarify with ATC before he can read it.

Sometimes, however, ATC will come right back with the full clearance; always have pencil in hand and be prepared to copy the clearance. In some cases there will be a delay between the time you call Clearance Delivery and when you actually receive the clearance.

UNACCEPTABLE CLEARANCES

If the clearance is unacceptable, read it back anyway, then explain. Valid reasons include:

- No survival gear for long over-water section.
- Icing conditions.
- Routing will add 50 minutes to the flight and exceed legal fuel reserves.

The pilot in command is responsible for the safety of the flight; any compromises with safety are unacceptable.

When operating from an airport with a control tower, ATC will issue a clearance on either a clearance delivery frequency or on ground control. If operating from an airport not served by a control tower (or if the tower is closed) there are several ways to get the clearance.

REMOTE COMMUNICATIONS OUTLETS (RCOs)

Check first to see if the uncontrolled departure airport has an RCO—a remote communications outlet. This transmitter and receiver antenna located on the airport is linked by landline to flight service or ATC. You can find RCOs listed in the *Airport/Facility Directory* in the communications section of the airport listing. (See FIG. 7-2.) Some RCOs are also shown on sectional charts.

Check the RCO listings carefully. Sometimes an RCO will receive on 122.1 MHz and transmit on a VOR frequency. There are quite a few RCOs scattered throughout the country at uncontrolled fields as well as some controlled fields where towers do not operate 24 hours a day.

If an RCO is located at an airport, request and copy a clearance and receive a "release" for takeoff.

CONNECTICUT

WINDSOR LOCKS

§   **BRADLEY INTL**   (BDL)   2.6 W   UTC–5(–4DT)   41°56′20″N 72°41′01″W   **NEW YORK**

      174   B   S4   **FUEL**   100LL, JET A   OX 1, 2, 3, 4   TPA—See Remarks   H-3D, 6J, L-25C, 28I

      LRA   ARFF Index D   **IAP**

      RWY 06-24: H9502X200 (ASPH–GRVD)   S-200, D-200, DT-350, DDT-710   HIRL CL

        RWY 06: ALSF2 TDZ. Trees.   **RWY 24:** MALSR, VASI(V4L)—GA 3.0°TCH 56′. Tree.

      RWY 15-33: H6846X200 (ASPH)   S-200, D-200, DT-350   HIRL

        RWY 15: REIL. VASI(V4L)—GA 3.5°TCH 59′. Trees.   **RWY 33:** MALSF. VASI(V4R)—GA 3.0°TCH 59′.

        Trees.

      RWY 01-19: H5246X100 (ASPH)   S-28   MIRL

        RWY 01: Building.   **RWY 19:** Trees.

      AIRPORT REMARKS: Attended continuously. Rwy 01–19 restricted to ldg and tkf with maximum tkf gross weight of
        73,000 pounds. This restriction does not apply to acft emergency. Numerous birds frequently on or in vicinity or
        arpt. Portions of taxiway Alpha not visible from tower. TPA—1174(1000) light acft, 1874(1700) heavy acft. Landing
        fee for business, corporate and revenue producing aircraft. Flight Notification Service (ADCUS) available.

      WEATHER DATA SOURCES: LLWAS.

      COMMUNICATIONS: ATIS 118.15   **UNICOM** 122.95

      BRIDGEPORT FSS (BDR) TF 1–800–WX–BRIEF. NOTAM FILE BDL.

      WINDSOR LOCKS RCO 122.3 (BRIDGEPORT FSS)

® **BRADLEY APP/DEP CON** 125.35 (260°–060°) 123.85 (061°–090°) 125.65 (091°–190°) 125.8 (191°–259°)

      TOWER 120.3   **GND CON** 121.9   **CLNC DEL** 121.75

      ARSA ctcAPP CON

      RADIO AIDS TO NAVIGATION: NOTAM FILE BDL.

      (T) **VORTACW** 109.0   BDL   Chan 27   41°56′27″N 72°41′21″W   at fld. 165/14W.

        VOR portion unusable 090°–103° beyond 24 NM below 5000′ 104°–170° beyond 10 NM below 6000′

        260°–290° beyond 15 NM below 6000′.

              DME portion unusable:

          040°–085°beyond 13 NM below 2000′   170°–195°beyond 14 NM below 3000′

          130°–150°beyond 10 NM below 3000′   250°–290°beyond 18 NM below 6000′

      CHUPP NDB (LOM) 388   BD   41°52′38″N 72°46′00″W   058° 5.2 NM to fld.

      ILS/DME 111.1 I-BDL Chan 48 Rwy 06 LOM CHUPP NDB

      ILS/DME 108.55 I-IKX Chan 22Y Rwy 33

      ILS/DME 111.1 I-MYQ Chan 48 Rwy 24

Fig. 7-2. *Typical listing in* Airport/Facilities Directory *for remote communications outlet (RCO).*

VOID TIME CLEARANCES

If there is no radio facility to issue an IFR clearance, it must be received by telephone. Clearances issued by telephone are called "void time" clearances because ATC will always set a time limit after which the clearance is void. When you file IFR by telephone from an uncontrolled field, ask whom to call to pick up a clearance. Flight service will provide a telephone number, which might be that flight service station's number or an ATC telephone number.

Call back 10 minutes prior to the proposed departure time. Flight service or ATC will read the clearance over the telephone and you will read it back. Make sure everything is ready to take off immediately. The clearance void time is usually 10 minutes after reception by telephone. If unable to take off within the time limit, contact ATC before the clearance is void and request a later "time window" for release and a new void time. The time window is the time between the release and the void time.

TAXI CHECKS

Three important checklist items should be emphasized on an instrument flight.

Turn Coordinator. How do the symbolic airplane and the ball move in a turn while taxiing? The "wings" tilt in the direction of the turn and the ball slides in the opposite direction to the outside of the turn. (On the older turn and slip indicator, the needle moves in the direction of the turn.)

The gyro of the turn needle is electrically powered. If the needle doesn't move, or moves erratically, it is either an instrument failure or an electrical failure. In either case, turn back. Your airplane is not safe for an instrument flight.

Attitude Indicator. The "wing" of the symbolic airplane should remain aligned with the horizon line. If the horizon display behind the symbolic wing pitches up or down or tilts beyond the slight movements seen during taxiing, it is an unreliable instrument. The gyros of the attitude indicator and the heading indicator are vacuum driven.

Cross-check the heading indicator. If it is drifting off heading, you probably have a vacuum system failure. Check the suction gauge. If it is not within limits—usually 4.6 to 5.4 inches of mercury—turn back because the airplane is not safe for flight. If the suction gauge reads normally and the attitude indicator is not showing a normal display, turn back.

Heading Indicator. A failure in the vacuum system will also affect the heading indicator. Headings shown on the heading indicator should change as you turn, then steady up to correctly match the magnetic compass, on long, straight taxiways. Cross-check with the attitude indicator and suction gauge. If they show abnormal indications, it is a failure in the vacuum system. If the attitude indicator and suction gauge are normal and the heading indicator is erratic, it is an instrument failure in the heading indicator.

ROLLING ENGINE RUN-UP

Normally those are the only checks made while the airplane is rolling. You do not want to get distracted while taxiing, particularly at night. There is an important exception to this rule, however, and that is when there is heavy commercial traffic sharing on the taxiways. The pilots of those big jets will certainly get upset if you come to a full stop and block the traffic for an engine run-up and takeoff checklist.

Do the run-up and takeoff checklist while taxiing. Practice this a few times with an instructor when the taxi traffic is light.

RUNWAY CHECKS

There are six items on the instrument runway check:

1. **Correct Approach Chart.** The departure airport's approach in use in case you have to return shortly after takeoff.
2. **Nav Radios Set.** Nav 1 for approach in use in case you have to return quickly; Nav 2 for first airborne fix.

3. **Correct Departure Control Frequency.** Be ready for a change to departure control quickly and without fumbling when tower hands you off.

4. **Lights On.** All on for night operations, strobes and rotating beacon on for day IFR flights.

5. **Transponder On and Set.** The correct code as given in the clearance.

6. **Pitot Heat On.** Even on a clear day get in the habit of doing this to avoid inadvertently overlooking this important safeguard when flying into visible moisture.

7. **Check Heading Indicator Against Runway Heading.** Sometimes the heading indicator will drift off while taxiing. Reset if necessary.

Just before adding power for takeoff, do an "STP" Check: Strobes, Transponder, Pitot heat.

And finally, note time of takeoff just before adding power. Takeoff time is doubly important on an instrument flight for fuel calculations and for planning the arrival in the event of lost communications.

IFR COMMUNICATIONS

Departure is the pilot's busiest time on an IFR flight: controlling the plane through a wide variety of situations, from parking and taxi through takeoff and into an instrument climb: transitioning between flight using outside visual references and flight by instruments: trying to maintain an efficient climb with turns to comply with departure clearance.

You must be listening for the airplane's N-number among many others tower may be working and be ready to respond quickly and correctly when tower has an amended clearance or issues a frequency change.

Reply to every ATC communication promptly. This is especially important during readback of an amended clearance. Read back immediately; if something puzzles you, figure it out later. ATC wants to hear from you right now. If a clearance is totally incomprehensible, say "stand by for readback" then call back and read back the entire clearance even if only the next section of the instructions is figured out. Research the rest of the clearance later.

STANDARD PHRASEOLOGY

Use concise, standard phraseology and a professional tone of voice when working with ATC and the controllers will be inclined to assist as much as they possibly can.

We've all heard transmissions like this: "This is Cessna November one two three four. I'm over (long pause), ah, Hartford! And I'm cruising at four thousand five hundred feet, departed from my homebase at, ah (pause) ten o'clock on a VFR flight plan to New Jersey. Request permission to go through the New York TCA. Over."

Do you think ATC will show any enthusiasm for clearing this clown through one of the country's busiest TCAs? Of course not. Time on the radio is very precious, especially around busy areas such as New York.

ATC is much more likely to grant a request if transmissions are brief and to the point. The more professional you sound, the more cooperation you will get from ATC. This is especially true with IFR communications. Appendix C, ''Pilot/Controller Glossary,'' contains many standard IFR phrases that should always be used as indicated. Frequently used terms are printed in bold italics.

WHO, WHO, WHAT, WHERE

Each time you switch to a new controller the initial call sequence consists of who, who, where, and what:

- WHO you are calling.
- WHO you are (always give the full N number on the first call).
- WHERE you are located (east ramp, outer marker inbound, Mist Intersection, etc.)
- WHAT your request or message is.

Use the full registration to avoid a possibility of confusing similar numbers, such as 62876 and 67876.

As a general rule abbreviate the call sign after the controller begins doing so, and use the controller's abbreviation. This speeds things up considerably. There is no need to say the ''N'' (November) for flights within the United States. ''N'' is the international designation for aircraft registered in the United States.

For example, if the full identification is N3456X, the initial call would be ''Cessna three four five six Xray.'' If the controller came back with ''Five six Xray turn right to zero nine zero,'' use ''five six Xray'' until switched to another controller.

If ATIS is available, be sure to copy the ATIS information before calling clearance delivery or ground control for taxi clearance. Give ground control the code for the ATIS and tell them that you are IFR. Ground control needs to know this so they can get a ''release'' from departure control or center. You cannot take off on an IFR flight until the controlling authority for the airspace issues a ''release'' to enter that airspace on takeoff.

CALLING GROUND CONTROL

The call to ground control might sound something like this:

''Westchester ground, Cessna three four five six Xray, terminal ramp with IFR clearance and information Delta. Ready to taxi.''

Delta would be the code for the current ATIS information. It is changed every hour or sooner if there is a significant change in safety information. It is important, especially in IFR weather, to let the ground controller know that you have the latest information. If you don't, the ground controller will request verification. This results in extra transmissions that clutter up the frequency and wastes everybody's time.

Say it on the first transmission and avoid using the phrase "with the numbers" instead of the ATIS code. All this phrase does is tell the ground controller that you either didn't bother to get the ATIS information, or that you got it and couldn't remember the code long enough to repeat it back.

You will remain on ground control until ready for takeoff, then switch to tower. On the first call to the tower controller give the full call sign: "Westchester Tower, Cessna three four five six Xray, ready for takeoff, IFR." Include "IFR" to alert the tower controller that the flight is IFR.

When switched to departure control, the next call might be: "New York Departure, Cessna three four five six Xray, out of one thousand for three thousand (1,000 to 3,000 feet)."

Nearly every first contact with a new controller requires acknowledgment of altitude or the altitude passing through to an assigned altitude. This helps the controller verify that actual altitude is the same as that reported by the Mode C transponder and shown on the controller's radar scope.

WHEN YOU HEAR NOTHING FURTHER

When one facility switches you over to another facility, it's a "handoff." What if it came time for a handoff—from tower to departure control, for example—and you received no further instructions? Wait and hope for something to happen? Call someone? If so, who? And say what?

This situation is not uncommon when the traffic is heavy. Sometimes the silence right after takeoff makes it feel like the tower has forgotten you.

If you have taken off and hear nothing further, continue climbing until at least 500 above the ground and established on departure heading. Then call the tower and ask: "Do you want four five Xray to go to departure control?" That will alert the tower to the fact that the handoff has not taken place.

Don't just sit there fat, dumb, and happy and say "Well, they didn't ask me to go to departure control, so I'm not going to do it." As the pilot-in-command, get things straightened out.

MANAGING FREQUENCIES

Many experienced pilots keep a running list of the various frequencies used during a flight so that it is easy to return to the last frequency in case the pilot or the controller makes a mistake in the next frequency. Another technique is to alternate between COMM 1 and COMM 2. Others like to work with only COMM 1 while en route, leaving COMM 2 tuned to the emergency frequency, 121.5 MHz with the volume adjusted to a comfortable level.

If, for some reason, no one answers after switching to a newly assigned frequency, simply go back to the last assigned frequency, give the abbreviated call sign, and say "unable" with the name and frequency of the facility you couldn't contact. For example the transmission might be:

"Five six Xray unable Boston Center one three three point one."

ATC will respond to an "unable" message with further instructions.

REQUIRED REPORTS

A major difference between VFR and IFR communications is that many IFR situations *require* reports to ATC. Pilots must make the following reports at all times, whether or not in radar contact with ATC:

- When leaving an assigned altitude.
- When changing altitude on a "VFR on top" clearance.
- When unable to climb or descend at a rate of at least 500 feet per minute.
- On commencing a missed approach. The report must include a request for clearance to make another approach attempt or to proceed to another airport.
- When average true airspeed at cruising altitude varies by more than 5 percent or 10 knots.
- On reaching a holding fix. The report must include time and altitude.
- When leaving a holding pattern.
- If, in controlled airspace, VOR, ADF, or ILS equipment malfunctions, or any impairment of air-to-ground communications.
- Any information relating to the safety of the flight.
- Any weather conditions that have not been forecast, or encountering hazardous conditions that have been forecast.

When not in radar contact (such as when ATC says "radar service terminated"), pilots must make these additional reports:

- En route position reports upon reaching all designated reporting points. These are indicated on en route charts by a solid triangle. Position reports include position, time, altitude, ETA and name of next reporting point, and name only of next succeeding reporting point. For example: "Cessna three four five six Xray, Carmel, one five, six thousand, Kingston two zero, Albany."

You don't even need to remember the sequence of items in the position report if the flight log form suggested in Chapter 3 is used. A reminder of the sequence is printed at the bottom of the left-hand page. (FIG. 3-1.)

- When leaving a final approach fix inbound on final approach.
- When it becomes apparent that an estimate previously given in a position report (above) is in error in excess of three minutes.

RADIO CONTACT LOST

Anticipate a handoff when you have been with one controller for quite some time and the controller requests "IDENT." Occasionally you may lose radio contact when flying at low altitudes or in areas of high terrain, or a long distance from the transmitter site. Call for a radio check; if there is no response after two or three attempts, try to re-establish communications through another nearby frequency listed on the IFR charts, such as flight service, approach control, or even a tower. As a last resort use the emergency frequency 121.5 MHz.

Always consider phraseology before transmitting and then speak in the most professional manner possible. Always be brief and to the point. Listen for a break in other transmissions and key the microphone immediately when a break occurs. Quickly give your call sign and message on the first call. That will save precious radio time for everyone on that frequency.

8
Basic Instruments

YOUR FIRST INSTRUMENT TRAINING FLIGHT WILL BE AN OVERVIEW OF THE CONDUCT OF an instrument flight—planning, weather, filing, departure, en route, approach, and landing—and it will expose what seems like an overwhelming amount of detail. However, a large amount of this detail is routine, procedural, and bears a close similarity to prior VFR planning.

By plunging right in at the beginning, you will soon find that even such "impossible" tasks as copying clearances will get easier and easier. This will foster attentive devotion to the basics of flying solely by reference to instruments, and to IFR instrument navigation, approaches, and the other important elements of instrument flight.

FIRST INSTRUMENT FLIGHT

When I take students up on their first instrument flight, I want them to be full participants in all elements of the flight. Of course I help the students when they find that something is a little too difficult to handle or the work load gets ahead of them.

But I want them to go through the motions throughout all phases of the flight—including a precision approach without the hood—so they can at least experience a visual picture of how it's done. It doesn't make any difference if the student makes mistakes because these are part of the learning process.

An amazing number of pilots find themselves able to take on a great share of the work load on this first training flight just by doing what comes naturally. This is especially true if the student has had good instruction in instrument fundamentals for the private pilot certificate.

However, he does stay within the parameters required for instrument flight by FAR 91. The student must stay within ± 100 feet of assigned altitudes and within the confines of whatever victor airways we use. I'm not looking for expert instrument flight this first time out. The objective on the first training flight is an introduction to the whole process, not a polished performance.

It has always surprised me how poorly the fundamentals of instrument flight are sometimes taught. All too often the student is told to fly straight and level for awhile under the hood and get the hang of it. The instructor continually criticizes, reminding the student that he is going off altitude or heading instead of emphasizing how altitude and heading are controlled.

The how-to is as important as the practice. Instructors should be able to show clearly how control is achieved and maintained. With the methods I have developed over the years, control of the airplane can be quickly fine-tuned to a level that is almost unbelievable.

TWO, TWO, AND TWENTY

From the very first flight I teach students to maintain heading within ±2°, airspeed within ±2 knots, and altitude within ±20 feet. This is necessary to become a member in good standing of the "2, 2, and 20 Club." This may sound corny, but it makes students think about goals to aim for right from the beginning.

The FAA allows tolerances on the instrument flight test of ±10° on heading, ±10 knots on airspeed, and ±100 feet on altitude.

Stop and think about it; these are really very large deviations—up to 20 degrees from one extreme to another on heading, 20 knots on airspeed, and 200 feet in altitude!

If a pilot is trained from the beginning to hold 2, 2, and 20, there will be no problem on the flight test. Even more important, he will have better control of the airplane. If the personal maximum altitude deviation is 20 feet, you will use much smaller control adjustments to maintain it, compared to a 100-foot deviation limit. Larger tolerances invite larger corrections and the larger the correction, the greater the tendency to overcontrol.

OVERCONTROLLING

Remember the problem from VFR training? Overcontrolling occurs when a pilot uses too large a change in attitude to make a correction—too much bank in a turn, nose too high or too low in a climb or descent, for example. Soon after the correction, the pilot corrects again in the other direction to avoid overshooting his heading or altitude. The flight of an overcontrolled airplane is a wobbling, bobbing affair in which it is impossible to hold headings and altitudes with any degree of accuracy.

Overcontrolling is, then, a problem of attitude control. And what is the best way to control attitude, avoid overcontrolling and achieve the goal of 2, 2, and 20?

Flight Lesson 2 concentrates on the fundamentals of controlling the aircraft by reference to instruments.

Instructor Note. The student should be under the hood as much as possible from here on. Raise the hood from time to time so the student can compare instrument indications to visual references.

It's helpful to remember that there are only four basic maneuvers in flying: straight and level, turns, climbs, and descents. Every move made with an airplane is based on these four maneuvers, or combinations of these four, such as a level turn.

ATTITUDE CONTROL

To make the aircraft perform these maneuvers the pilot must maintain or change its attitude. The attitude indicator was specifically developed as the basic reference for maintaining or changing attitude in instrument flight. The other flight instruments are vital, of course, and lifesavers if the attitude indicator should fail. But for normal IFR flying, the attitude indicator is the star player, and the other flight instruments are the supporting cast.

Straight and Level

I have found it best to work first on straight and level flight. Most of the time an instrument cross-country is straight and level flight. And, more often than not, the other basic maneuvers begin and end with a return to straight and level.

This basic maneuver is divided into two elements: altitude control and heading control. It seems to work better to practice altitude control first, then work on heading control.

Choose an odd or even thousand-foot altitude—such as 3,000, 4,000, or 5,000 feet—for practicing straight and level. It's interesting to note in passing that on an IFR flight, most of the time you will be flying at an odd or even thousand foot altitude with the long hand of the altimeter pointing straight up to zero.

Set the power for cruise and trim the airplane. Trim is important—the closer you can trim the airplane to fly "hands off," the easier it will be to make the small corrections that are so vital to precise instrument flight.

ALTITUDE CONTROL

In level flight, the representative wings on the attitude indicator should superimpose the horizon line to form one line (See FIG. 8-1). If it doesn't look like this in level flight, reset the attitude indicator.

Try maintaining altitude within ±20 feet for 20 seconds with reference to the attitude indicator primarily, using the altimeter as a cross-check. If altitude drops, use slight back pressure on the control yoke to pitch the nose up slightly. Raise the wings of the miniature airplane to one-half bar width above the horizon line, keeping the wings level (FIG. 8-2). (Bar width is the thickness of the wings of the miniature airplane represented on the attitude indicator.) When you get back to the desired altitude, relax the back pressure and align the wings with the horizon line.

Fig. 8-1. *In straight and level flight, the miniature airplane on the attitude indicator should be set so that its "wings" make a straight line from left to right with the horizon line.*

Fig. 8-2. *When below altitude by 100 feet or less, correct by increasing pitch one-half bar width.*

Fig. 8-3. *When above altitude by 100 feet or less, correct by decreasing pitch one-half bar width.*

If altitude increases, apply a slight forward *pressure* on the yoke to pitch the nose down slightly. Lower the miniature airplane to one-half bar width below the horizon line, keeping the wings level (FIG 8-3). Cross-check with the altimeter. Level off at the desired altitude. Relax the forward pressure on the yoke and align the wings with the horizon line again. *Pressure* (not movement) is the key to smooth control.

Step Climbs and Descents

The ''step'' climb and descent is particularly good for developing an automatic reflex for making altitude corrections; it is very helpful to the student in reaching the ''20'' part of the 2, 2, and 20 goal in climbs and descents and in controlling altitude.

See FIG. 8-4, which shows the maneuver from a starting altitude of 3,000 feet.

Fig. 8-4. *Step climbs and descents for practicing altitude adjustments.*

Instructor Note. The step climb and descent is very useful for another reason: It can be introduced early in the syllabus without overloading the student.

After a student has demonstrated an ability to hold some convenient altitude consistently for two or three minutes on a constant heading, practice the step maneuver next. He should climb 100 feet, level off, and hold that altitude for one minute. The miniature airplane on the attitude indicator is held one-half bar width above the horizon line with back pressure alone (FIG. 8-2). No power changes are needed for a change of less than 100 feet.

After the minute is up, descend to the original altitude, level off, and hold that for one minute. The miniature airplane should be pitched one bar width below the horizon line for the descent (FIG. 8-3). Again, there should be no change in the power setting.

One-half bar width on the attitude indicator is a very small adjustment, but with a little practice, it can be easily achieved every time. One-half bar width is also an excellent rule of thumb: For altitude changes of 100 feet or less, use no more than a one-half bar width correction.

The remainder of the exercise consists of one more descent and one more climb performed the same way—establishing pitch with the attitude indication and making no power changes. The wings should remain level according to the attitude indicator.

The exercise should be continued until altitude can be controlled within ± 20 feet of what is desired. Students who master step climbs and descents usually have little difficulty with altitude changes of more than 100 feet.

CONTROL, PRIMARY, AND SUPPORT INSTRUMENTS

The attitude indicator provides pitch information during the step maneuver; it is the central control instrument or simply the control instrument.

Primary information on the quality of control comes from the altimeter. It shows a direct, almost instantaneous reading on whether or not the quality is good enough to stay within ± 20 feet.

Here's a good way to determine the primary instrument: It indicates the most pertinent information about how well you are doing and *does not move* when flying precisely. The altimeter is the primary instrument in maintaining level flight. Does it meet the test? You bet it does. It indicates the most pertinent information regarding altitude. And if you're doing a good job, the altimeter will not move.

The vertical speed indicator (VSI) has neither a control nor a primary role maintaining level flight. But it is an additional source of information on the rate of climb or descent and may be used for cross-checking. The VSI is thus a support instrument during straight and level flight.

FIGURE 8-5 shows the primary, support, and control instruments for all phases of instrument flight. Note how the instruments just discussed are used to control pitch in level flight. I will classify instruments as control, primary, and support when discussing other instrument flight maneuvers. Refer to this chart from time to time to better understand the roles these instruments play during each phase of flight. The primary instruments will have added importance later during partial panel practice because they become control instruments during simulated failure of the vacuum-driven attitude and heading indicators.

	CONTROL	**PRIMARY**	**SUPPORT**
STRAIGHT and LEVEL			
Pitch	Attitude Indicator	Altimeter	VSI (Rate of Climb)
Bank	Attitude Indicator	Heading Indicator	Turn Coordinator
Power		Airspeed	RPM/MP
SPEED CHANGES			
Pitch	Attitude Indicator	Altimeter	VSI
Bank	Attitude Indicator	Heading Indicator	Turn Coordinator
Power		Airspeed	RPM/MP
STANDARD RATE TURN			
Pitch	Attitude Indicator	Altimeter	VSI
Bank	Attitude Indicator	Turn Coordinator	Sweep Second Hand
Power		Airspeed	RPM/MP
MINIMUM CONTROLLABLE AIRSPEED			
Pitch	Attitude Indicator	Airspeed	RPM/MP
Bank	Attitude Indicator	Heading Indicator	Turn Coordinator
Power		Altimeter	VSI

Fig. 8-5. *Basic Instruments: Control, primary, and support instruments for all the basic regimes of flight.*

Fig. 8-5. Continued.

CLIMB ENTRY

Pitch	Attitude Indicator	Attitude Indicator	VSI
Bank	Attitude Indicator	Heading Indicator	Turn Coordinator
Power		RPM/MP	

CONSTANT AIRSPEED CLIMB

Pitch	Attitude Indicator	Airspeed	VSI
Bank	Attitude Indicator	Heading Indicator	Turn Coordinator
Power		RPM/MP	

CONSTANT RATE CLIMB

Pitch	Attitude Indicator	VSI	Altimeter, Sweep Second Hand
Bank	Attitude Indicator	Heading Indicator	Turn Coordinator
Power		Airspeed	RPM/MP

LEVEL OFF (to cruise from climb or descent)

Pitch	Attitude Indicator	Altimeter	VSI
Bank	Attitude Indicator	Heading Indicator	Turn Coordinator
Power		Airspeed	RPM/MP

CONSTANT AIRSPEED DESCENT

Pitch	Attitude Indicator	Airspeed	VSI
Bank	Attitude Indicator	Heading Indicator	Turn Coordinator
Power		RPM/MP	

CONSTANT RATE DESCENT

Pitch	Attitude Indicator	VSI	Altimeter Sweep Second Hand
Bank	Attitude Indicator	Heading Indicator	Turn Coordinator
Power		Airspeed	RPM/MP

CONTROL	=	Main reference instrument
PRIMARY	=	Key quality instrument*
SUPPORT	=	Back-up or secondary instrument

Variable Power (cruise, etc.) - "Power to the speed, pitch to the altitude."
Constant Power (min. controllable airspeed, etc.) - "Pitch to the speed, power to the altitude."
Power + Attitude = Performance
*The Primary Instrument is always the instrument that gives the most pertinent information and is not moving when flying precisely.

If you can maintain altitude within $\pm$ 20 feet for 20 seconds, we'll try it for 40 seconds, then carry it on to the next step, one minute. Once you get a good grasp of altitude control with this method I will cover up the altimeter, wait 30 seconds or so, then uncover it to see how far up or down we have drifted. With a little practice you will be amazed at how well you can fly without looking at the altimeter. This also serves as painless little introduction to partial panel work, which we will take up on later flights.

Instructor Note. Once again it is a good idea to raise the hood every now and then so the student can compare what he sees on the instruments with the real horizon and other visual references. For example, when the nose is pitched up on the attitude indicator, raise the hood and show him that the nose is also pitched up with reference to the actual horizon. This sounds obvious but it's not obvious to the beginning instrument student unless actually pointed out. The beginning instrument student needs all the help he can get at this point!

Before moving on to turns, climbs, and descents, let's work on straight and level until it becomes automatic. Concentrating on one element at a time, you will succeed in learning very quickly to work within tolerances of 2, 2, and 20. The control pressures necessary to maintain altitude in straight and level flight are minimal, the tendency to overcontrol is much reduced, and even the heading will remain relatively constant because wings are level. And best of all, you will soon do this without hardly thinking about it!

Most readers recall getting a bicycle for a birthday or Christmas. Within the first week or two you probably wore the skin off your knees learning to ride by trial and error. Then after two weeks, suddenly the secret of balancing came and you were in full control. You could command that bicycle to turn left or right merely by leaning or shifting weight on the seat of the bicycle. And it wasn't long before you were riding along gleefully in front of friends with arms folded, just leaning left and right to make turns.

Turning had become an automatic reflex. Not reacting, but acting. That's the secret of straight and level flight.

Students develop an automatic reflex so that if the long needle of the altimeter is to the left of the big zero, low, apply back pressure to move the miniature airplane one-half bar width above the horizon line. And if the long needle is to the right, apply forward pressure to lower the miniature airplane one bar width below the horizon line.

I practice this with a student until he is making corrections automatically for sudden updrafts and downdrafts without pausing to analyze what needs to be done. Then it will be time to move on to heading control. This isn't the end of altitude control. I'll have much more to say when considering climbs and descents to accomplish altitude changes.

HEADING CONTROL

As an expert at maintaining altitude in straight and level flight, it's time to excercise the same finesse to control and maintain heading.

Set up straight and level on a convenient heading. With wings level on the attitude indicator, apply control pressure equally and very gently to both rudder pedals. It's not necessary to be heavy-footed; just keep both feet in contact with both pedals and balance the pressure.

Constant heading with wings level is maintained by small adjustments in the balance of pressure on the rudder pedals. Without constant pressure on the rudder pedals, the airplane will drift off heading if it yaws due to poorly adjusted rigging or constantly changing air currents.

Make minor adjustments in the pressure on the rudder pedals to compensate for heading changes caused by these factors. For heading adjustments of 5° or fewer use rudder pressure only, holding wings level; flight will be uncoordinated momentarily, but you will avoid the tendency to overcontrol.

Overcontrolling is an impulse that all students seem to have and they work consciously to avoid it. Think about it, if you want to make a two degree change of heading and you roll in 20 degrees of bank, what happens? Suddenly the heading has changed 20 or 30 degrees, and you are powerfully tempted to roll into a steep bank in the opposite direction.

As the plane swings back and forth through the sky, control deteriorates and precise flight becomes impossible. When you were a student pilot, your instructor probably took over at this point and steadied the plane so the lesson could continue. But now you are expected to re-establish control without a helping hand. Or much better still, use very small control inputs to avoid overcontrolling altogether.

If the plane tends to drift left with wings level, increase the pressure on the right pedal. If it drifts right, increase the pressure on the left pedal. In other words, "step on the ball" to make minor corrections with rudder pedals. As learned in VFR training, apply rudder pressure in the direction the ball has moved to return to coordinated flight.

If the heading has drifted off more than 5° make a coordinated turn to adjust the heading, using an angle of bank that is *no more than half* the desired degrees of heading change. If you need to adjust your heading 10°, for example, put in an angle of bank of no more than 5°. This will help avoid overcontrolling.

While airplane rigging and changing air currents are common causes of heading drift, the major culprit is banking. The attitude indicator, once again, is the control instrument for keeping the wings level. If you maintain a carefully adjusted pressure on the rudder pedals, the wings will stay level almost automatically. The heading indicator is the primary instrument that gives a direct, almost instantaneous quality reading of efforts to maintain level flight because every change in bank will result in a heading change.

The supporting instrument for level flight is the turn coordinator. The turn needle is extremely sensitive and will often show evidence of a bank that is almost imperceptible on the attitude indicator. The ball must remain centered with rudder pressure in level flight or else the airplane will yaw and drift off heading.

SCAN

While practicing altitude and heading control in straight and level flight by means of control, primary, and support instruments an efficient scanning techique will develop. It's something that happens almost automatically as you work with attitude, altitude, and heading in straight and level flight. The scan develops even further to include the support instruments.

Students often ask what is the best way to scan an instrument panel: up and down, left to right, clockwise or counterclockwise? It doesn't really make much difference as long as you adhere to two very important principles: (1) don't get fixated on one instrument, and (2) always return to the attitude indicator, the control instrument, after checking each of the other instruments.

Teaching Scan

I have found that an ideal method of teaching the scan is *attitude, heading, altitude.* Repeat these three words while flying to help guide your scan.

Look first at the attitude indicator to determine if any attitude corrections are required. Next, check the heading indicator to see if any drift in heading has occurred. Glance at the altimeter to determine if any correction in altitude control is required. Go back to the attitude indicator and make any adjustments dictated by the glance at heading and altitude.

Repeat this process continually throughout the flight. On every 10th scan, include the VOR or the ADF, depending on which is used for navigation. Every couple of minutes include all the instruments in your scan, including engine gauges. The main thing is to adopt a system and stick with it.

DISTRACTIONS

The human eye is constructed to immediately respond to any movement it picks up. This is a well-known trait from ancient forebears who had to respond quickly to movement of any kind in their environment; movement either indicated the presence of something to eat, or the presence of something that would eat them.

The implication for the modern instrument pilot is that we descendants still tend to respond with greatest interest to something that moves. Our attention tends to become fixed on the oscillations of the turn needle or the rapid up-and-down movement of the VSI needle in bumpy weather instead of the attitude indicator. Our attention is riveted on the wanderings of the VOR needle near a station while neglecting heading and altitude.

Make an effort to tear your eyes away from an instrument that is showing rapidly changing indications and methodically scan from one instrument to another in whatever sequence is most comfortable.

Refer to FIG. 8-5. The control instrument for all phases of instrument flight is the attitude indicator. Center the scan pattern on the attitude indicator to maintain control of the airplane on instruments. When eyes move away from the flight instruments—setting a new communications or navigation frequency—always return to the attitude indicator to resume the scan.

You will quickly see small changes in attitude develop and make the small, automatic adjustments of yoke and rudder pedals that are so important in controlling the airplane.

SUPPORT INSTRUMENTS

Now, let's look at the role of support instruments, which serve two purposes. They may be required to assume the role of *control* instruments in case of mechanical failure. Should the gyro in the attitude indicator fail, the airspeed indicator can now substitute for the attitude indicator, supported by the VSI. (At a certain power setting you know what the airspeed should be in level flight. A five knot increase in airspeed would indicate a slight descent and vice versa.) The ADF needle (tuned to a strong standard broadcast station) in combination with the magnetic compass could substitute for the heading indicator, and so on.

The second purpose of the support instruments is to stabilize flight and minimize overcontrolling and to detect instrument malfunctions. If the airspeed remains steady on the indicator, for example, and your VSI and altimeter both show a rapid descent, you can conclude that the airspeed indicator is malfunctioning.

AIRSPEED CONTROL

Achievement of precise altitude and heading for longer and longer periods in straight and level flight leads to precise airspeed control.

Recall from VFR flying that to increase airspeed in straight and level flight it is necessary to increase power: to decrease airspeed, reduce power. The same principles apply to IFR flying, of course, but applied more precisely than before.

Precision will demand very small changes in power settings—increments of 100 rpm or 1 ″ of manifold pressure in a high performance airplane, unless a specific power setting is required. Light control pressure made it easier to hold altitude and heading.

Likewise small power changes will also make it easier to control the airplane within the goal of 2, 2, and 20. Large changes in power, or "throttle jockeying," is a form of overcontrolling and must be eliminated or else precise instrument flight cannot be attained.

A power change of 100 rpm or 1 ″ of manifold pressure yields a change in airspeed of 7½ knots. Strangely enough, this rough rule of thumb can be applied to any propeller driven airplane you are likely to fly, Cessna 152, Cherokee Arrow, Beech Baron, or whatever. For example, if you are cruising at 100 knots and want to reduce your airspeed to 95 knots, the 5 knot change would equate to a reduction of 75 rpm or ¾ ″ manifold pressure. To increase or decrease airspeed airspeed by 10 knots, you would increase or decrease rpms by 125 or change manifold pressure by 1¼ ″.

Remember that this rule of thumb is not set in concrete. Minor adjustments are almost always necessary. It is something to start with. I will offer many rules of thumb throughout this book; they are valuable guides to precision IFR flight and you won't have to waste time reinventing them. (Don't waste time trying to rediscover old knowledge!)

AIRSPEED TRANSITIONS

Let's take our discussion of airspeed control one step further and talk about airspeed transitions—how to change from one airspeed to another with facility and precision.

First, establish a specific speed for each flight condition for the type of airplane flown. For a Cessna 172 in level flight, for example, consider 110 knots as normal cruise, 90 knots as slow cruise, and 70 knots as slow flight.

Next, establish the power settings to maintain these selected speeds in a Cessna 172. Maintain 110 knots with 2,450 rpm, 90 knots with 2,100 rpm, and 70 knots with 1,900 rpm.

Practice airspeed changes while flying under the hood.

To slow down from 110 to 90 knots, for example, reduce power to the setting established for that speed, 2,100 rpm.

Maintain level flight by slowly raising the nose of the miniature airplane on the attitude indicator, with wings level. Increase back pressure on the yoke as the airspeed bleeds off, and trim out the pressure as you approach the desired airspeed.

Cross-check with the primary instruments: altimeter for the best information on pitch, heading indicator for bank, and airspeed for power. Include the support instruments in your scan: VSI for pitch, turn coordinator for bank, and tachometer for power.

To speed up and resume 110 knots, reverse the procedure. Increase power to 2,450 rpm, and maintain level flight by slowly lowering the nose with reference to the horizon line on the attitude indicator. Increase forward pressure on the yoke, then trim out the pressure near 110 knots. Cross-check the same primary and support instruments.

Power changes induce airspeed changes, while adjusting attitude changes altitude. In straight and level cruising flight it's "power to the speed, pitch to the altitude." You'll hear instructors say this over and over as you make airspeed changes in level flight.

POWER

There are only two basic power conditions for flight: variable (adjustable) and not variable (either by choice or by accident).

You use variable power in level flight to control airspeed (power to the speed, pitch to the altitude). Variable power is also used in climbs and descents at a specified rate (such as 500 feet per minute) and on the glide path of an ILS approach. You will see how this works later when these maneuvers are analyzed.

The conditions under which power is not variable occur when using full throttle on takeoff, when the throttle is closed or the engine fails, or when the power is in transit during a transition from one maneuver to another, such as intercepting a glide slope.

In these situations remember "pitch to the speed, power to the altitude" from VFR training and flight at minimum controllable airspeed.

INSTRUMENTS THAT LIE—WHEN AND WHY

Mr. Murphy had a great gift for simplifying technical problems in an unforgettable way. His first law: "If a part can be installed wrong, it will be."

Murphy's law that applies to instrument flying: "If an instrument can fail, it will."

Here, in summary form, is a list of instruments and problems to prepare for.

Attitude Indicator. This instrument is driven by gyros powered by the vacuum system. The indicator fails when there is a failure in the vacuum system, the result, usually, of an engine-driven vacuum pump failure. The heading indicator may also fail even when the vacuum system is operating normally.

Failure of the attitude indicator shows up gradually. The instrument doesn't just roll over and die but begins to drift off slowly at first as its gyros wind down. For a good idea of what this looks like, watch the way an attitude indicator behaves after engine shut down at the end of a flight.

If you have trouble maintaining straight and level, stabilized climbs and descents, or smooth standard rate turns, suspect failure of the attitude indicator. Cover it up to prevent scan distraction and switch to partial panel operation, which is explained in Chapter 12.

The attitude indicator may also show erroneous information if it has been set incorrectly. There is only one situation in which the attitude indicator can be set correctly. And that is straight and level unaccelerated flight. That is the only time the miniature airplane can be accurately matched to the horizon line.

If you reset the attitude indicator while sitting on the flight line, it will be incorrect due to the nose up attitude of the parked airplane. Other errors will occur if a reset is attempted in turns, climbs, and descents.

Heading Indicator. Like the attitude indicator, the heading indicator is driven by the vacuum system. It will fail if the vacuum system fails, and it can also fail independently. Like the attitude indicator, a failure of the heading indicator is rarely dramatic. If you suspect a failure, cover the instrument and switch to partial panel operation.

Make sure that the heading indicator shows the same heading as the magnetic compass. Students make three common mistakes when setting the heading indicator.

First, they fail to reset the heading indicator when lined up on the runway centerline for takeoff. This is the best time to reset a heading indicator because you know the runway heading. If it's an instrument runway, the heading is often shown to the last degree on an approach chart instead of rounded to the nearest 10° interval.

Second, because of precession the heading indicator slowly drifts off. It should be reset every 15 minutes, or after maneuvers that involve a lot of turns in a short time, such as holding patterns.

Third, the heading indicator cannot be reset accurately unless the airplane is in straight and level, unaccelerated flight. The magnetic compass is the culprit here. It is accurate only when stabilized in straight and level, unaccelerated flight. (Compass errors and how to deal with them will be detailed in Chapter 12.) Meanwhile, resist the temptation to reset the heading indicator in a turn. I have seen students make errors of as much as 30 degrees while trying to match heading indicator and magnetic compass in a turn.

Magnetic Compass. I will have a lot more to say about the magnetic compass in Chapter 12, but let's review a few points learned in VFR training:

1. In calculating headings, account for *variation* due to the Earth's magnetic field, and *deviation* due to magnetic influences on a specific compass because of its location in the airplane.

2. Turning toward the north, the compass *lags* behind the turn due to dip error, turning toward the south, the compass *leads* the turn.

3. On easterly and westerly headings, acceleration produces an indication to the north, deceleration produces an indication to the south. Remember ANDS (Acceleration North Deceleration South).

Altimeter. The altimeter will read erroneously if not set to the correct barometric pressure at all times. It will also read erroneously if the static port is clogged. Insects, ice, and dirt can clog a static port. The problem becomes apparent when airborne and the altimeter needles don't move.

An alternate static source aboard an airplane can restore the altimeter to normal operation if the static port has become clogged. However, the altimeter will read *higher* than normal.

If you do not have an alternate static source aboard, create one by breaking the glass face of the VSI. This vents the pitot-static system to the cabin, the same as an alternate static source. Again, the altimeter will read higher than normal. (Breaking this glass face of the VSI usually damages the needle and renders the instrument inoperative.)

Airspeed Indicator. A blockage of the pitot tube will render the airspeed indicator useless. Like the altimeter, you won't know this until airborne.

Pitot heat will prevent ice from clogging the pitot tube. That's why I recommend turning on the pitot heat before takeoff on an instrument flight. When ice has clogged the tube, pitot heat may melt it too slowly.

Insects love to nest in the pitot tube; use the pitot tube cover. Insects can clog those tiny air passages in just a few minutes.

Blockage of the static source also causes erroneous airspeed readings. Use of an alternate static source will produce an indicated airspeed a knot or two *higher* than normal.

Vertical Speed Indicator (VSI). The most important thing to remember about the VSI is that it only gives an accurate reading when the needle has been stabilized for seven seconds or longer. If the needle is moving, forget it.

Another quirk of the VSI is that when you first raise the nose of the airplane to begin a climb, the VSI needle initially shows a descent. The reverse is true in the beginning of a descent when the needle will momentarily show a climb.

The needle of the VSI should point to zero when the aircraft is sitting on the ground. If not, the needle can usually be zeroed by turning a small screw at the lower left corner of the instrument case. If this adjustment cannot be made, add or subtract the error for an accurate reading in flight.

Turn Coordinator. The turn coordinator is powered by electricity so it will continue to operate even if there is a failure of the vacuum powered attitude indicator and heading indicator. If the turn coordinator fails, the needle won't move. It will remain fixed in an upright position. That's why it's important to check the movement of the turn coordinator while taxiing out.

It's pretty rare, but I have also seen the ball of the turn coordinator get stuck in the tube.

Fuel Gauges. Here's another Murphy's law: On land, air, and sea, the second half of the tank always empties faster than the first half!

Never trust any fuel gauge. There is no way to judge how accurate they are and most of them are fairly crude. Always note your time of takeoff and the time en route from each major position fix, then calculate fuel consumption based on airplane performance figures.

Ask "What if my fuel gauges failed completely? Am I keeping track of fuel calculations well enough independently of the gauges to know exactly how much more flying time I have left?"

Oil Pressure and Temperature Gauges. Engine instruments should be scanned every few minutes. The main concerns are low oil pressure and rising oil temperature. When these symptoms appear, a serious problem is developing in the engine oil system. Land the airplane as soon as possible. Don't stop to think about whether or not the gauges are functioning properly.

Low oil pressure with no rising oil temperature indicates either an instrument error or an incorrectly set pressure relief valve. Keep an eye on this situation. As long as the indications remain stable, the flight may be continued. But if the oil pressure drops and the oil temperature rises, land.

High oil pressure with normal oil temperature usually means that the pressure relief valve has been set incorrectly. As long as engine indications remain normal there is no reason to discontinue the flight.

9
Turns, Climbs, and Descents

AS A VFR PILOT YOU BECAME ACCUSTOMED TO FLYING STABLE, MEDIUM BANKED TURNS. The basic turn used for instrument flying—a standard rate turn—is much milder, requiring about 15 degrees of bank in a Cessna 172 at cruising speed. It is easy to execute, doesn't require a lot of attention during the turn, and altitude control is less demanding than a steeper turn.

Overcontrolling is a common problem. The habit of setting up that steeper VFR medium banked turn could be so deeply ingrained that you might bank too steeply on the first practice standard rate turns. If you bank too steeply, the rate of turn increases rapidly. Then you have to reduce the angle of bank to slow everything down. And by that time the altitude is way off. Take it easy!

Standard rate turns are precision turns that produce a heading change of exactly 3° per second. If the maneuver starts out on a heading of 360°, then becomes a standard rate turn to the right, the heading in 30 seconds will be 090°, in one minute 180°, and so forth.

Enter the turn by using the attitude indicator to establish the angle of bank for a turn rate of 3° per second. The formula for the degree of bank is

$$\frac{\text{Airspeed}}{10} + 5 = \text{bank angle for a standard rate turn.}$$

Example: Airspeed is 125 knots, divided by 10, plus 5, equals a 17.5° bank angle for a standard rate turn.

TURN COORDINATOR AND CLOCK

The turn coordinator is the primary, or quality, instrument. In a standard rate turn, the symbolic airplane tilts and its wings line up with left and right benchmarks (FIG. 9-1). The ball remains centered throughout the turn with proper rudder input. (In standard rate turns with an older turn and slip indicator, the needle is displaced one needle width in the direction of the turn.)

The sweep-second hand of the clock may be included in the scan during standard rate turns. It is an invaluable support instrument because it shows whether you're turning faster or slower than 3° per second.

No matter where the second hand is when you commence a turn, it should make half a sweep for every 90° of turn, and a complete sweep for every 180° of turn.

If you have turned 90° and the second hand is more than halfway around the clock face, you know your rate of turn is too slow. Speed up the rate of turn by increasing the angle of bank slightly. Likewise, if you have turned 90° and the second hand has not reached halfway, you know your rate of turn is too fast. Decrease the rate of turn by decreasing the angle of bank slightly.

Fig. 9-1. *Standard rate turn to the left at 100 knots. The wings of the miniature airplane match the benchmark on the turn coordinator. But the angle of bank falls between marks on the attitude indicator because the airspeed is 100 knots.*

Be careful, however, not to fixate on the clock. That steadily advancing sweep-second hand is a powerful attention getter. Make a conscious effort to direct attention away from the clock and return to your normal scan, resuming, as usual, with the attitude indicator.

Practice level standard rate turns in both directions for a full 360°, then try turns of 180° and 90°. To hit the rollout heading exactly, anticipate the target heading by one half the degrees of bank, just like VFR turns to a heading. If the angle of bank in the turn is 20°, for example, begin the rollout 10° before reaching the target heading. A 10° angle of bank requires a lead of 5°.

TRIM IN A TURN

A level turn requires back pressure on the control wheel to maintain altitude. Students frequently ask me whether or not they should apply nose-up trim during a turn to reduce the back pressure needed to keep the nose up. This is an individual matter and there is no strict rule for trimming. I think you will find that a little extra back pressure on the yoke during the turn will work very well in counteracting the tendency of the nose to descend. But it's really up to you; find the method that works best for you and stick with it.

Cross-check attitude indicator, altimeter, and VSI to make sure you are using the right amount of back pressure to maintain constant altitude. In a full 360° turn, which requires two minutes, it *is* helpful to add a quarter of a turn of nose-up trim. Remember, though, that this trim will have to be removed after completing the turn or the nose will rise.

OBOE PATTERN

Practice rolling smoothly out of a standard rate turn in one direction, without pausing, directly into the opposite direction. Here I would like to introduce the Oboe pattern, which is designed to help perfect standard rate turn technique, plus rolling from one direction to another (FIG. 9-2).

Start from straight and level flight on a cardinal heading with the sweep-second hand of the clock approaching 12 o'clock. Begin a standard rate turn to the left and continue the turn for 360° (two minutes). Include the clock in the scan so that you can adjust the rate of turn according to whether it is slow or fast passing the 6 and 12 o'clock positions.

The original cardinal heading becomes the ''reversal'' heading. Use the ''one-half the angle of bank'' rule of thumb to anticipate the reversal and turn to a standard rate turn in the opposite direction.

Don't pause in straight and level flight at the reversal heading; roll smoothly into the turn in the opposite direction. The nose will tend to pitch up during the maneuver. Prepare for this and apply forward pressure on the yoke. The attitude indicator is a big help during the turn reversal; just keep the dot between the miniature wings aimed right at the horizon line and your altitude will remain constant. Think in terms of riveting that dot on the horizon and rolling around it during the reversal.

I think you will enjoy flying the Oboe pattern, especially if you feel a bump from wake turbulence indicating a perfect circle as you come around to the original cardinal

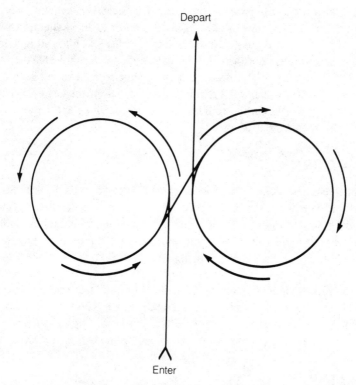

Depart

Enter

Fig. 9-2. *Oboe Pattern. Complete the first 360° turn and roll right into the second 360 without pausing when wings are level.*

heading. Do the pattern with both left and right entries. The Oboe will reappear when it is time to practice climbing and descending turns and partial panel procedures.

PATTERN A

It's time to assemble all the elements covered so far: straight and level, speed changes, and standard rate turns. A good exercise for this is Pattern "A", shown in FIG. 9-3. It also contains all the maneuvers required for a full-scale instrument approach, except for the descents. You will fly this pattern, or portions of it, on every instrument approach.

Make a copy of Pattern "A" and attach it to your clipboard as a ready reference.

Set up the exercise in straight and level flight at normal cruise speed. Start the pattern on 360° the first few times you try it. As soon as you are comfortable with the pattern, vary the initial headings and start on 090°, 180°, or 270°. This will be good practice to learn which rollout headings will be with different initial headings.

Start timing at the beginning of the exercise with the sweep-second hand at 12 o'clock. Time each leg consecutively; each new leg starts when the time for the old leg has expired and control pressure is applied to adjust for the new leg.

Fig. 9-3. *Pattern A.*

Both the ''Oboe'' Pattern and Pattern ''A'' are excellent exercises to practice under the hood without an instructor, but you must have an ''appropriately rated'' safety pilot in the right seat. (See Appendix B and FAR 91.21 (b)(1).)

Don't continue practicing these patterns if problems develop. It might be frustrating and you might unknowingly develop bad habits in scanning or procedures. As any instructor will tell you, it takes a lot of extra time and effort to break bad habits. Try again another day with an instructor.

MINIMUM CONTROLLABLE AIRSPEED

I like to introduce minimum controllable airspeed under the hood early in the training syllabus. This surprises a lot of students. They are just beginning to master straight and level flight under the hood and I ask them to try flight at minimum controllable airspeed!

Introducing minimum controllable airspeed early serves several important purposes. First, it will improve your handling ability on instruments because you will learn to control the aircraft throughout a wide range of pitch and power changes. And you will learn to anticipate the major changes in control pressures that accompany these transitions. A student who learns to handle the plane well in minimum controllable airspeed under the hood, always finds it easier to reach the goal of 2, 2, 20 when flying straight and level and other fundamental maneuvers.

Second, the student learns to extend and retract flaps (and landing gear if so equipped) while under the hood. With very little practice extending and retracting flaps and gear on instruments becomes an automatic reflex. This is very important on instrument approaches. The last thing you want to happen while descending on final is to break your scan while fumbling with flaps and landing gear.

Simulation of Missed Approach

The third reason for practicing minimum controllable airspeed under the hood also has to do with instrument approaches. Imagine this situation: You have flown a perfect approach down to the ceiling minimum and the runway is not visible; it's time to execute a missed approach. The airplane is low and slow, the nose is pitched up, and flaps are extended. To stop the descent, you must add full power, level off, and start to retract the flaps.

The airplane might be as low as 200 feet above the runway. You must be in absolute, positive control of the airplane to increase airspeed rapidly and establish a climb to avoid obstructions and put some distance between yourself and the ground.

The transitions during recovery from minimum controllable airspeed are identical to those during a missed approach: full power, increase speed, retract flaps, maintain altitude.

Missed approaches will be introduced and practiced in later lessons. Minimum controllable airspeed provides several fundamental building blocks that must be mastered early in the course to become a precise instrument pilot who can handle all maneuvers with absolute safety.

The best way to master minimum controllable airspeed on instruments is to think that it is an extension of the second fundamental item, speed changes, except that the changes are carried to greater limits.

Entering the Maneuver

Set up the maneuver from straight and level flight on a convenient cardinal heading. Start reducing power; for the Cessna 172 or Piper Cherokee 180, you will find that reducing power to 1,500 rpm will be adequate. As the nose starts to feel heavy, and the airspeed slows to the flap extension range, apply the first notch of flaps, or 10°. As airspeed decreases further, lower the flaps another notch. As the nose gets heavy again, extend full flaps. Do not attempt to drop full flaps all at once because the changes in attitude and control pressures will be so abrupt and heavy that you will have difficulty controlling the airplane by reference to instruments.

As minimum controllable airspeed approaches, the instructor will request power applications (perhaps full power) to maintain altitude at the lowest possible speed without stalling. Aim for the airspeed at which altitude control becomes difficult. That will be the minimum controllable airspeed for this exercise, and it will not place the aircraft in an imminent stall situation.

Another important lesson to be learned from minimum controllable airspeed is when flying with full power, the power is "constant" and not "variable." So you will have to "pitch to the speed, power to the altitude," as we discussed earlier in the section on airspeed changes.

In other words, maintain airspeed by adjusting nose attitude and maintain altitude by adjusting power. Once again, the attitude indicator is the instrument used to control the airplane. The airspeed indicator is the primary, or "quality" instrument for pitch. The heading indicator is primary for bank and the altimeter is primary for power (FIG. 8-5). Include VSI, turn coordinator, and tachometer or manifold pressure in your scan as supporting instruments for pitch, bank, and power, respectively.

Instructor Note. The student must do clearing turns prior to minimum controllable airspeed. Make sure there are no airplanes in the area because the high nose-up attitude in this maneuver will limit your vision as safety pilot. My recommendation is that in the beginning the student should do standard rate turns left and right 90° prior to reducing power.

As the student gains more experience, the clearing turns may be made while reducing power and lowering flaps. This is good practice because in the real world of IFR there will be many times during approaches when the plane will have to be slowed and flaps extended while in a turn.

CONSTANT AIRSPEED CLIMBS

In VFR flying you have already become accustomed to constant airspeed climbs. After take off, climb at a speed that will give the best rate of climb, usually 80 knots for a Cessna 172 or Cherokee 180. In IFR flying, just as in VFR, the most important thing is to pick the best rate of climb speed (V_y) and stick to it.

Full power is used for climbs, so climbs are situations in which power is "constant," not "variable" and you will "pitch to the speed."

Climb speed is thus established by setting the pitch attitude. You will set and control the pitch attitude by reference to the attitude indicator.

Establish the climb by pitching up to that first index line in the blue "sky" section of the attitude indicator (FIG. 9-4). If entering the climb from straight and level, add full power and simultaneously increase pitch.

Fig. 9-4. *Pitch up to first index line above horizon line on attitude indicator to establish climb.*

Pitch Adjustments

Make adjustments from that first reference line, as necessary. If airspeed is too high and climb performance is poor, correct by a slight additional pitch up adjustment to reduce the airspeed; if airspeed is too low, make a slight pitch down adjustment. Hold the pitch constant on the attitude indicator and the airspeed will remain constant.

Pitch adjustments on the attitude indicator—like all adjustments in instrument flight—are small and deliberate, not more than a quarter or half a bar width at most. Bar width is the thickness of the miniature airplane "wings" on the attitude indicator. The smaller the adjustments, the smoother the flight.

Note that when entering a climb, the attitude indicator serves a triple purpose: the control instrument and the primary instrument for pitch and the control instrument for bank. The main concern in climb entry is establishing the correct pitch; the attitude indicator gives the best information on the "quality" of pitch, as well as serving as the reference by which you control the airplane in both pitch and bank.

When the climb is stabilized at the best rate of climb speed, the airspeed indicator becomes the "quality" instrument for pitch. The primary and support instruments for both climb entry and stabilized climb are the same, as seen in FIG. 8-5.

DEPARTURE CLIMBOUT

Be prepared on the first instrument training flight to climb out on instruments after takeoff. Under ATC clearance, adhere to standard instrument climbs, turns, and descents. Make full use of this excellent opportunity for practicing instrument climbs and you won't have to use valuable time later in the flight to hone climb skills.

Let's see how this works in the real world of IFR. The clearance from ATC contains climb instructions; ATC frequently requires a step or two in the climbout before reaching cruise altitude for better traffic separation. A typical clearance might be:

"Maintain three thousand feet, expect further clearance to five thousand feet in ten minutes."

In this clearance there are two climbs. First is the initial climb to 3,000 feet after takeoff. Then there is a stretch of level flight until ATC calls back 10 minutes later with further clearance to climb to 5,000 feet. How, in practical terms, do you handle this type of climb clearance?

As a rule of thumb, climb at the highest practical rate up to the last 1,000 feet before the assigned altitude. Then reduce the rate of climb to a constant 500 feet-per-minute for the last 1,000 feet to avoid overshooting the assigned altitude when leveling off.

In the previous example, you would take off and climb to 2,000 feet with full power at the best rate of climb airspeed. Then you would adjust the attitude to produce a 500 foot-per-minute climb from 2,000 to 3,000 feet. Then level off until ATC clearance to resume the climb to 5,000 feet, approximately 10 minutes later.

CONSTANT RATE CLIMBS

On reaching 2,000 feet you would switch from a stabilized *constant speed* climb to a stabilized *constant rate* climb at the rate of 500 feet per minute. (You will find that this is less complicated in flight than it sounds in print.)

Most single engine light planes can't manage much more than a 500 foot-per-minute climb at full throttle anyway. To attain a constant rate climb, simply reduce pitch slightly, leaving the throttle at full power.

The VSI is the primary instrument for "quality" information on the rate of climb. See FIG. 8-5 again. You will have to wait approximately seven seconds for the VSI to stabilize at the new pitch. In constant rate climbs and descents the attitude indicator continues to be the instrument by which you control the airplane. The VSI becomes the primary instrument for attitude. See FIG. 8-5 once again.

If the VSI stabilizes at more than 500 feet per minute, lower the nose to decrease the rate of climb; if less than 500 feet per minute, raise the nose to increase the rate of climb. Always remember that you cannot get an accurate indication from the VSI until it has stabilized for seven seconds.

CHASING THE NEEDLE

When I give instrument check rides I can tell very quickly if candidate pilots do not understand the limits of the VSI because they "chase the needle." If the needle is rising fast, they push forward on the yoke to slow it down; if it is descending fast they pull back. This produces a form of porpoising, which is a sure tip-off that the student has neither developed a good scan nor an understanding of instrument control. Instead his attention remains fixed on one instrument too long.

Instructor Note. A student who develops the VSI needle chasing symptom has an instrument fixation problem. Return to level flight and teach the student to force himself to shift focus from one instrument to another, starting and stopping each scan cycle with the attitude indicator.

CLIMB LEVEL OFF

Leveling off from a climb is a simple procedure, but it's the source of a common problem for beginning instrument students. That problem is overshooting or undershooting the target altitude. To avoid this, anticipate reaching the target altitude by 10 percent of the rate of climb. If the VSI shows a rate of climb of 600 feet per minute, for example, begin leveling off 60 feet before the target altitude.

Lower the miniature airplane to the horizon line on the attitude indicator when reaching the desired altitude and allow the airspeed to build up to cruise before reducing power. Exert forward pressure on the yoke to prevent ballooning above the assigned altitude as airspeed builds up, and trim out the excess control pressure as it becomes heavier.

It is very important to have a predetermined idea of what the cruise power setting should be. You can automatically reduce power to that specific cruise power setting in one motion when you reach cruise speed. Any further adjustments of power and trim will then be relatively minor.

Include the sweep-second hand of the clock in your scan. No matter when you start the constant rate climb, you know that you should gain 500 feet by the time the sweep-second hand returns to its starting position. The altitude gain should be 250 feet when the sweep-second hand is "30 seconds" from its starting position.

DESCENTS

Overshooting the target altitude when descending is a more critical matter. Most precision instrument approaches go down to 200 feet above the runway, where there's no room for error. You must be able to control descents so there is never a question of coming too close to the ground or obstacles in the airport vicinity.

It's worth noting that airspeed control during descents becomes very important when making VOR and other nonprecision approaches. Understand that the Missed Approach Point (MAP) on a nonprecision approach is frequently based on how long it takes at a given airspeed to fly from the final approach fix to the airport.

If you can't control your airspeed on a descending final approach course, you might as well throw the stopwatch away. Elapsed time on that final leg to the airport will be meaningless if the airspeed on which it is calculated is not constant. If the time is off, you may end up off course. If there are hills or obstacles around the airport, poor timing due to poor airspeed control might be disastrous.

CONSTANT SPEED DESCENTS

In a constant speed descent use "pitch to the airspeed" like the constant speed climb. While in straight and level flight, reduce power to slow cruise. Set 1,900 rpm in a Cessna 172 or Cherokee 180, approximately 90 knots for a comfortable and efficient descent speed.

"Pitch to the airspeed" again. Hold the nose up in level flight until speed bleeds off to 90 knots, then gently allow the nose to pitch down to one line below the horizon line

Fig. 9-5. *Pitch down to first index line below horizon line on attitude indicator to establish descent.*

on the attitude indicator (FIG. 9-5). That should produce a 90-knot descent at about 500 feet per minute. (It may vary slightly for different types of airplanes.)

"Pull the plug" if you need to descend more rapidly. Reduce power to 1,500 rpm, slow to 90 knots, then lower the nose to maintain 90 knots. This will produce a rate of descent of about 1,000 feet per minute at a somewhat steeper angle.

Some high performance airplanes can descend 2,000 feet per minute by reducing power and lowering the landing gear and flaps. The "dirty" descent (gears and flaps down) can be made without reducing power excessively. This procedure avoids the shock of sudden overcooling, which can damage the engine.

Constant Rate Descents

Let it be known here and forever carved in stone that a change of 100 rpm or 1″ of manifold pressure will produce a change of 100 feet per minute (fpm) in the rate of descent. Suppose you are in a 400-feet-per-minute descent with power at 1,900 rpm, and airspeed 90 knots. How do you change the rate of descent to 500 feet per minute? Simply reduce power by 100 rpm to 1,800. Power adjustments of 100 rpm or 1″ in descents will produce changes of 100 feet per minute in almost every plane you are likely to fly.

This rule of 100 rpm or 1″ = 100 fpm will become vitally important later on as you learn to fly the ILS approach with its very sensitive electronic glide slope. But the rule also applies to normal descents, and it will help avoid overshooting that critical altitude.

ATC will expect a descent at 1,000 fpm until reaching 1,000 feet above the target altitude. Then you are expected to reduce the rate of descent to 500 fpm. In other words, a change to a lower altitude begins with a *constant speed* descent at 90 knots and approximately 1,000 fpm. At 1,000 feet above the target altitude, change to a *constant rate* descent at 500 fpm.

This is similar, of course, to the situation in a climb. The first part of a climb is at a *constant speed*—the best rate of climb speed. At 1,000 feet below the target altitude it changes to a *constant rate* climb of 500 fpm. The first part of a descent is at a *constant speed.* At 1,000 feet above the target altitude, you change to a *constant rate* descent of 500 fpm.

Refer again to FIG. 8-5 and to see that the instruments have the same control, primary, and support roles in a descent as they do in a climb. Again the sweep-second hand of the clock should be included in the scan to ascertain whether you are ahead or behind in the climb.

DESCENT LEVEL OFF

The target descent altitude should be anticipated by 10 percent of the rate of descent— the same as in a climb. If descending at a rate of 500 fpm, begin to level off 50 feet prior to reaching the target altitude.

Instrument students sometimes fixate on the target altitude and don't start the transition until they reach it. That is a sure way to overshoot the altitude and, on an instrument approach, to go below the minimums for that approach. Don't do it! ''Busting the minimums'' is hazardous to your health when low to the ground. And on an instrument check flight, busting the minimums is an automatic failure, no matter how brilliant the rest of the check ride might have been.

To level off, simultaneously increase power to the cruise setting and raise the nose and set it on the horizon bar of the attitude indicator. Adjust trim as the speed builds up to prevent ballooning above the level-off altitude. Scan the attitude indicator and the primary and support instruments as shown in FIG. 8-5.

APPROACH DESCENTS

Before leaving the subject of descents, I want to cover two slightly modified descents used on instrument approaches. (I will spend more time on these two descents in the approach phase of the training syllabus, but you should understand what they are at this point and see how closely they relate to standard descents.)

The first is a constant airspeed descent with the addition of 10° of flaps. Let's call them approach flaps because they may be more or fewer than 10° in different airplanes. Most airplanes handle better at slow speeds with a notch of flaps down. This is the configuration used on most instrument approaches on the final approach leg just before landing.

Start from cruise speed, reduce power, and extend approach flaps. Allow the airspeed to bleed off while in level flight, then ''pitch to the airspeed.'' Set the nose of the miniature airplane on the first black line below the horizon on the attitude indicator. Adjust power for 500 fpm, remembering that 100 rpm or 1″ = 100 fpm, or ''power to the altitude.''

If flying a high performance airplane with retractable landing gear, reduce power, set approach flaps, and allow the speed to decrease to desired descent airspeed while remaining in level flight. When you reach the descent speed, extend the gear; the nose dips automatically to just about the right pitch for a descent when the gear is lowered.

HIGH SPEED FINAL

More and more these days ATC might say: "Keep your speed up on final." This is frequently followed by something interesting like "DC-9 overtaking."

If you cannot comply with this request, do not hesitate to tell ATC right away. But it is best to cooperate with this request whenever possible, for obvious reasons. ATC won't let the separation between you and that big jet get too narrow. If a jet is behind you coming in 10 or 20 knots faster than you and chewing up the distance in between, guess who is most likely to be ordered to go around, you or the big jet? ATC might request a 90° left or right turn to let the jet pass before vectoring you back for the approach.

Practice constant rate descents at cruise speed, or higher, as well as at the normal descent speed. Some pilots automatically make high-speed final approaches whenever they fly into airports with a lot of jet traffic. This certainly makes it easier for ATC.

To set up a high speed descent, lower the nose of the miniature airplane to the first black line below the horizon line on the attitude indicator. Reduce power 500 rpm to set up a 500 fpm descent. When stabilized, "pitch to the airspeed, power to the altitude" to maintain cruise airspeed at a 500 fpm descent.

Leave flaps and landing gear up on a high speed approach in a light plane. If jets are on the approach behind you, you know the runway will be plenty long enough to slow down and extend flaps and gear when the runway is in sight and landing is assured.

VERTICAL S

When students file and depart IFR on every flight, they will usually have more than enough opportunity to practice climbs and descents in the real world of IFR. So there isn't much point in practicing additional vertical maneuvers. An instructor has to avoid a natural tendency to teach mechanics of the maneuver rather than the goal of the maneuver.

The Vertical S (FIG. 9-6) and its variations, the S-1 and S-2, are excellent exercises for an instrument student to practice with a safety pilot. The Vertical S consists of climbs to 500, 400, 300, and 200 feet with reversals at the top of each climb and descents back to the original altitude before climbing to the next altitude in the series. The Vertical S can also be a series of descents as shown in FIG. 9-6.

The Vertical S-1 is a combination of the Vertical S and a standard rate turn. Make a standard rate turn each time you return to the original altitude. Alternate turns to the left and to the right.

The Vertical S-2 differs from the S-1 in that the direction of turn is reversed with each reversal of vertical direction.

PATTERN B

The Vertical S, S-1, and S-2 are recommended maneuvers in the FAA's *Instrument Flying Handbook* (see "The Instrument Pilot's Reference Library"). However, I have found that Pattern B is much more effective in teaching students how to combine the

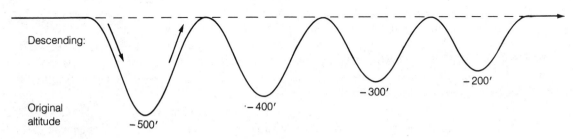

Fig. 9-6. *Vertical S practice maneuver.*

fundamentals of instrument flight: straight and level, speed changes, standard rate turns, climbs, and descents. It's an excellent maneuver for "putting it all together."

As you see in FIG. 9-7, the turns and straight stretches in Pattern "B" are the same as those in Pattern A. But B adds speed changes, and it includes a descent and an emergency pull-up to simulate an approach and missed approach.

Roll out on headings regardless of time passage. The turn to the final leg is a descending standard rate turn. Note that a pre-landing check is included, then a little later you extend ¼ flaps as to commence the final "approach." If you are flying an airplane with retractable gear, also lower the landing gear at this point.

At the emergency pull up, don't forget to retract approach flaps and landing gear, if so equipped. Does something seem familiar here? Right! It's the recovery from minimum controllable airspeed. The pieces indeed come together at this point. Maybe not perfectly, but the goal of 2, 2, and 20 is in sight.

PATTERN C

Don't worry, Pattern C isn't required! But you will feel a great sense of achievement if you can do it. It has been called a basic airwork "graduation exercise." (FIG. 9-8.)

If you can fly C with its non-standard climbs and descents and keep to 2, 2, and 20, you will have certainly mastered the fundamentals of attitude instrument flight. Patterns

Fig. 9-7. *Pattern B.*

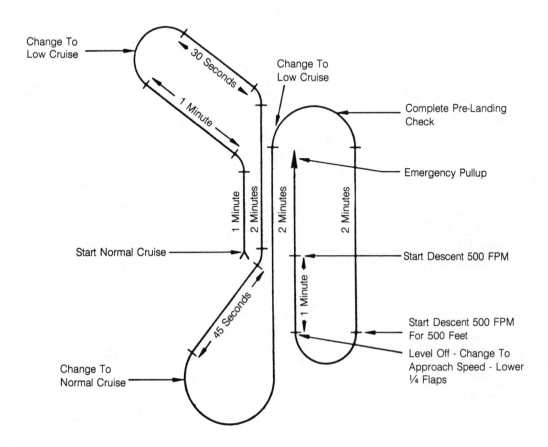

"B" and "C" are good exercises to practice with a safety pilot. Break off practice if the pattern work is not going well, otherwise you may unconsciously develop bad habits. Work with an instructor on whatever is causing the problem before any bad habits have a chance to take hold.

Master Pattern C on full panel, then try it on partial-panel. It is a sure cure for overconfidence; it is also instant insanity. Some dedicated instrument students have done this. I think they were former military pilots who had partial-panel Pattern Cs inflicted on them by sadistic military instructors. Civilian instructors, of course, would never pull a stunt like that. Flying is supposed to be fun!

Fig. 9-8. *Pattern C.*

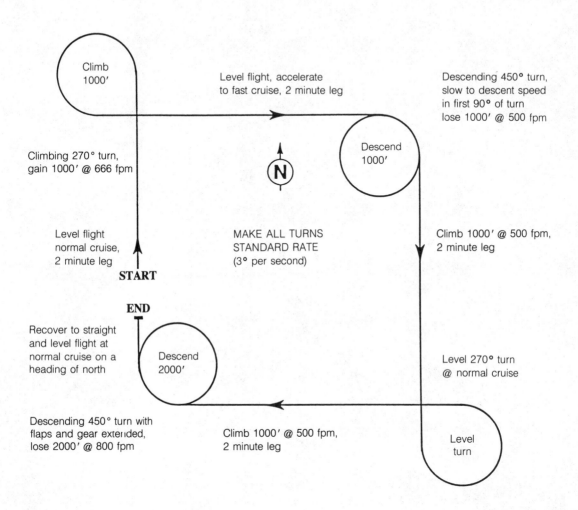

Climb 1000'

Level flight, accelerate to fast cruise, 2 minute leg

Descending 450° turn, slow to descent speed in first 90° of turn lose 1000' @ 500 fpm

Climbing 270° turn, gain 1000' @ 666 fpm

Descend 1000'

N

Level flight normal cruise, 2 minute leg

MAKE ALL TURNS STANDARD RATE (3° per second)

Climb 1000' @ 500 fpm, 2 minute leg

START

END

Recover to straight and level flight at normal cruise on a heading of north

Descend 2000'

Level 270° turn @ normal cruise

Descending 450° turn with flaps and gear extended, lose 2000' @ 800 fpm

Climb 1000' @ 500 fpm, 2 minute leg

Level turn

10
VOR Procedures

THE VHF OMNIDIRECTIONAL RANGE STATIONS (VORs) ARE THE HEART OF THE AIRWAY system. The radials from a VOR are highways in the sky. VOR radials form most intersections, and VORs are used for more instrument approaches than any other type of facility.

VOR use requires positive identification of the facility. Obviously it can be fatal if you fail to make a positive identification and use the wrong frequency on an instrument approach. So get in the habit of automatically turning up the volume and checking the identification for every VOR.

HEADING INDICATOR ERRORS

Another point to emphasize is periodically checking the heading indicator and readjusting it to match the magnetic compass. Maintaining an accurate course is difficult if not impossible if the heading indicator has drifted off.

The gyro of the heading indicator may precess a small amount due to bearing friction. The turns and reversals during instrument departures, approaches, and training maneuvers will produce additional precession errors.

A heading indicator that precesses up to 3° every 15 minutes is within acceptable tolerances. I recommend checking the heading indicator at least every 15 minutes and prior to intercepting the final approach course on all approaches.

Several other times the heading indicator must be checked or reset:

- When you line up on the center line of the active runway for takeoff. This is the best opportunity to set the heading indicator with greatest precision: The gyro is up to speed, the plane is stable, and you know the runway magnetic course.

- When you begin every approach, even after a missed approach. This item must be on the approach check list.
- On leaving a holding pattern and after practicing Patterns A, B, and C, holding, or any similar maneuvers that require numerous turns.
- After practicing unusual attitudes. These maneuvers may cause the heading indicator to wander off considerably because of the extra bearing friction produced by the maneuvers.

VOR PROFICIENCY

In the beginning of an instrument student's training, VOR skills must be determined. Does the student understand the basic principles of VOR orientation, and intercepting, tracking, and bracketing bearings and radials? Have any bad habits crept in? For example, does the student tend to get fixated on the course deviation indicator (CDI) needle and neglect other instrument indications? Remember, the goal remains 2, 2, and 20 in VOR work as well as other phases of instrument flight.

Don't spend expensive flight time to determine VOR proficiency. Make a quick pencil-and-paper check of the basic principles of VOR orientation by completing the VOR diagnostic exercise in FIG. 10-1.

Begin at the top line with the omni bearing selector (OBS) set at 030°. Then pick the correct VOR presentation for each lettered position. For example, with the OBS set at 030°, the most appropriate display for the A position is number 5. Proceed across through the G position, then drop down and complete the 090° line the same way, and so on.

Answers are in FIG. 10-2 on the next page. If any answer is wrong, review the exercise with an instructor. Proceed to the exercise shown in FIG. 10-3 for more VOR fundamentals.

16-POINT ORIENTATION

I developed this exercise many years ago. It works very well when teaching private pilot students exactly what happens around a VOR. The 16-Point Orientation Exercise is both a teaching exercise and a good diagnostic exercise because it will quickly reveal whether or not a student understands the basic principles of VOR. If not, the exercise can be repeated to bring the student up to par in short order.

Flying from west to east, with a VOR station to the south, one setting of the OBS will reveal 16 lines of position with precision while flying around the VOR station.

Most students are puzzled when I describe this exercise the first time. Sixteen lines of position from a single OBS setting? How can this be? It's really very simple when you understand what's going on, and when you understand, you will have mastered the basic principles of VOR work.

See FIG. 10-3. Enter the pattern on a heading of 090° with a convenient VOR station to the southeast. Set the OBS to 090°. As you begin the exercise, the TO/FROM indicator will show TO.

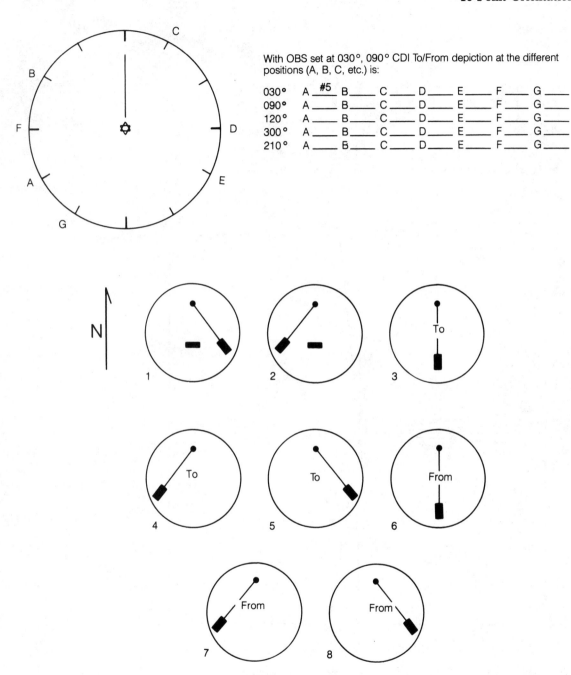

With OBS set at 030°, 090° CDI To/From depiction at the different positions (A, B, C, etc.) is:

030°	A _#5_	B ___	C ___	D ___	E ___	F ___	G ___
090°	A ___	B ___	C ___	D ___	E ___	F ___	G ___
120°	A ___	B ___	C ___	D ___	E ___	F ___	G ___
300°	A ___	B ___	C ___	D ___	E ___	F ___	G ___
210°	A ___	B ___	C ___	D ___	E ___	F ___	G ___

Fig. 10-1. *VOR diagnostic test. Answers on the next page.*

VOR Procedures

**With OBS
Set at:
CDI/To-From Depiction:**

030°	A **#5**	B **#1**	C **#6**	D **#7**	E **#2**	F **#5**	G **#3**
090°	A **#4**	B **#5**	C **#8**	D **#6**	E **#7**	F **#3**	G **#4**
120°	A **#4**	B **#3**	C **#1**	D **#8**	E **#6**	F **#4**	G **#2**
300°	A **#8**	B **#6**	C **#2**	D **#4**	E **#3**	F **#8**	G **#1**
210°	A **#7**	B **#2**	C **#3**	D **#5**	E **#1**	F **#8**	G **#7**

Fig. 10-2. *Answers to VOR diagnostic test on previous page.*

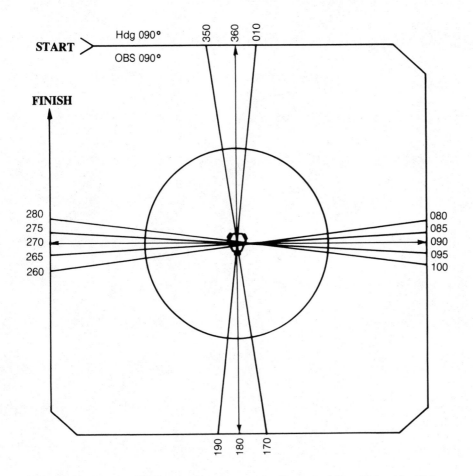

Fig. 10-3. *16-point VOR orientation exercise. Begin heading 090 with OBS set on 090. You get 16 lines of position without changing OBS setting.*

The 360° radial is the boundary between TO and FROM. The red flag will start to appear approximately at the 350° radial. When the red flag indicates OFF, you will be passing the 360° radial. That's the first precise line of position. When the red flag disappears, you will be approximately on the 010° radial. So the first leg of the exercise gives at least one precise line of position at 360° and 350° and 010° with lesser accuracy.

Continue on the 090° heading another two minutes or so, then turn right to a heading of 180°. The CDI needle will soon come alive and start moving in from right to left. When it reaches the outermost dot, you will be on the 080° radial. When the needle reaches the edge of the bull's eye in the center, you will be at the 085° radial, and when the needle centers you will be at least 090° radial. That's three lines of position.

As you continue on the 180° heading, the needle will pass the other edge of the bull's eye at 095°, and the last dot on the left at 100°, adding two more lines of position.

You have produced three lines of position on the first leg and five on the second leg for a total of eight. If you continue the pattern as shown in FIG. 10-3, you will add three more lines of position on the heading of 270°, and five more on the 360° heading for a grand total of 16 for the full exercise—all without resetting the OBS!

If further practice is needed, enter the maneuver at other cardinal headings.

Instructor Note. Have the student anticipate and call out the radials while flying around the VOR station. This will help the student visualize what is going on and it will also give you a good indication of whether or not the student understands what is happening.

VOR TIME/DISTANCE CHECK

Another excellent exercise in teaching the fundamentals of VOR orientation is the VOR Time/Distance Check, or poor man's DME. Use this procedure to estimate how long it will take to reach a VOR station in no-wind conditions. And, to tell the truth, I have never heard of a situation where someone has had to use the Time/Distance check to determine the time to a VOR on an actual flight. However, practicing the Time/Distance Check will sharpen VOR skills, and, as we shall see later, introduce you to a similar procedure used in making a DME arc approach.

Turn toward a convenient VOR station and adjust the OBS so the CDI needle centers in the configuration. Note the heading to the station.

Next, turn the airplane 80 degrees right or left of the inbound course. Rotate the OBS in the *opposite* direction of the turn to the nearest 10° increment. In other words, if you turn right 80°, turn the OBS *left* (counterclockwise) to the nearest 10° increment (FIG. 10-4). You are flying a short tangent to an imaginary circle around the station.

Maintain the new heading. When the CDI needle centers, note the time. Continue on the same heading and change the OBS another 10 degrees in the same direction as above. Note the number of seconds it takes for the CDI needle to center again. Divide the number of seconds by 10 to determine the time to the station in minutes. The formula is:

Fig. 10-4. *VOR time-distance check.*

$$\text{Minutes to station} = \frac{\text{Time in seconds}}{\text{Degrees of bearing change}}$$

You can also calculate the distance to the station by using this formula:

$$\text{Distance to station} \quad \frac{\text{TAS} \times \text{minutes flown}}{\text{Degrees of bearing change}}$$

For the second leg turn 20° *toward* the station and stop the turn. Turn the OBS 10° in the *same* direction as the turn. When the needle centers, note the time and turn another 10° in the same direction. When the needle centers again, note the elapsed time and calculate time to station.

Turn the airplane another 20° toward the station to start a new leg, and repeat the process. Continue these short tangents around the station as many times as you wish.

Complete the exercise by selecting an inbound bearing to the station that lies ahead of the last tangent leg you plan to fly. Start turning inbound 10° before reaching the inbound course.

When are you 10° from the inbound course? Simple. Start turning inbound when the needle reaches the outermost dot on the CDI display. This will indicate that you are 10° away from the inbound course.

If you need more practice in either the 16-Point Orientation Exercise or the Time /Distance Check, use a simulator. You can get in more practice in an hour in a simulator because you don't have to copy clearances, take off, and fly to a practice area. Whenever you begin to have difficulty, you can stop the exercise and analyze what's causing the problem. Furthermore, simulator time is less costly and more fuel efficient than airplane time.

INTERCEPTING A BEARING OR RADIAL

The first practical application of VOR work will most likely be clearance to a VOR station after takeoff: cleared "direct" to the first VOR on the clearance or to intercept a specific radial.

Clearance via a specific radial involves intercepting that radial and tracking inbound to the station on it. Two assumptions will simplify this procedure.

First, departure instructions and any radar vectors from ATC will always point in the right direction to make a quick, efficient intercept, unless you are being diverted away from traffic or higher terrain. This is also true with en route clearance changes that direct you to intercept a specific radial.

Second, clearances always state *radials*, not magnetic courses to the VOR. Because radials radiate *from* the station, you will fly *to* the station on the reciprocals of the radial. A quick and easy way to determine the reciprocal is to refer to the heading indicator. Follow the radial from its number on the edge of the dial, through the center, and out in a straight line to the number on the opposite outer edge (FIG. 10-5).

Set the assigned radial on the OBS and continue on the assigned heading. When the needle begins to move, you are on the radial that lies 10 degrees before the target radial.

Fig. 10-5. *Use heading indicator to visualize reciprocals.*

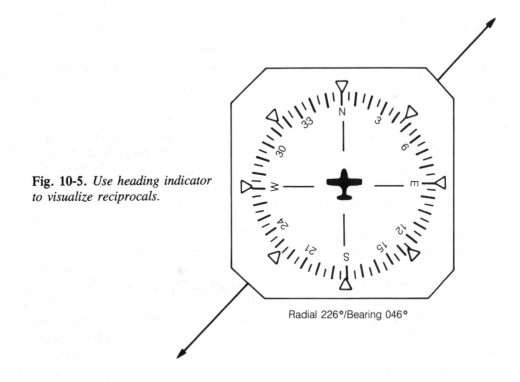

Radial 226°/Bearing 046°

Inbound Turn

When you reach this lead 10° radial, turn to intercept the inbound bearing at an angle of 60 degrees. When the needle reaches half-scale deflection, turn an additional 30° and maintain this 30° intercept heading until just before the needle reaches the bull's eye, the small center circle on the CDI presentation. When the needle touches the bull's eye, set the OBS to your inbound magnetic course.

See FIG. 10-6 for an illustration of how this works. In this situation, you have been assigned a heading of 320° and you have been cleared to the VOR via the 226° radial. As you reach the lead radial—216°—the needle comes alive. Turn right to a heading of 346° to set up a 60° intercept angle.

If you're not sure what heading will produce a 60° angle, refer again to the heading indicator, and count off 60° from the inbound bearing (FIG. 4-13). You will see this is 346°. With a little practice, you will be able to read reciprocals and intercept angles off the heading indicator at a glance.

When the needle reaches half-scale deflection, turn right again to 016°. Hold 016° until the needle reaches the bull's eye, then steady up on the inbound heading of 046°, plus or minus whatever wind correction is necessary to hold the needle in the center.

WIND CORRECTIONS

I'm always surprised when instrument students are unable to offer even an educated guess when asked, "Which way is the wind coming from?" Knowledge of the wind should almost be second nature by the time a person receives a private pilot certificate. If not, work on it during instrument training. An instructor should keep asking "which way is the wind?" until you begin anticipating and adjusting for the wind automatically.

Flight planning revealed wind forecasts at various points and altitudes and you know exactly what the wind was at takeoff. It is a simple matter of deciding whether the wind is going to push to the left or to the right departing the airport toward the first VOR fix.

A tailwind will speed interception and a headwind will delay interception. And when turning onto the inbound heading, add a wind correction factor automatically, maybe 2°, 5°, or 10° according to your best estimate. Refine this correction by making adjustments en route toward the station.

There are two other interception techniques. If you believe you are close to the station when intercepting a radial, make the first turn 45° toward the station, rather than 60°. Hold that 45° interception course until the needle is about three quarters of the way from full-scale deflection.

The needle reaches this position about 2½° from the assigned radial; turn to the inbound heading at this point. Add a correction for the wind when established on the inbound course.

How can you tell if you're close to the station? The more sensitive the needle, the closer you are to the station.

When in close, the 45° intercept will put you on the inbound bearing quickly and

You are at point "A"

B. As soon as the needle starts to move, turn to 346° (60° Intercept Angle).

C. When needle reaches half scale deflection, turn further right to 016°.

D. Join the 226° radial.

Note: Alternate method is to use 045° angle at "B".

Fig. 10-6. *Steps to intercept a VOR radial.*

closer to the staion. This extra distance will provide time to adjust the inbound heading for the wind, also providing a better chance of being exactly on course over the station.

The other interception procedure is a re-interception technique utilized when off course and the needle is pegged at full deflection. This happens when a strong wind changes abruptly or when you are seriously distracted and drift left or right without correcting the problem.

In either case, make an en route correction to return to the desired radial or bearing. If the wind is from behind, use a small intercept angle of 10 or 20 degrees to return to course and avoid overshooting; if a headwind, use an intercept angle of 20° or 30°. The larger angles will get you back on course sooner.

COMMON INTERCEPTION MISTAKES

The Needle Never Centers. This indicates that (1) you turned to the inbound heading too abruptly or too soon, or (2) a headwind was much stronger than anticipated. In either case, use the re-interception technique described above. Turn 10°, 20°, or 30° toward the needle and wait until the needle centers to resume the inbound heading.

The Needle Passes Through the Center and Moves Toward the Opposite Side. As in the first case, the inbound heading turn may have been faulty. You may have made the turn too slowly or waited too long to start the turn or the tailwind was stronger than anticipated. Use the re-interception technique to center the needle.

Another possibility is that you were so close to the VOR that the width of that 10° arc from full deflection to the center may have been only a few feet.

A 45° intercept angle one mile from the VOR is almost impossible. The needle will peg with a FROM indication almost as soon as you turn to the inbound course. Ideally, have at least five miles before you get to the VOR to do a skillful job of intercepting a radial. The only solution is to steady up on the outbound heading and re-intercept after the needle has settled down in the FROM position.

A good interception with the needle perfectly centered after the first turn to the inbound heading is not a matter of luck. With practice, and using the correct techniques, you will learn to judge the wind and turn so the needle will center every time.

CLEARED DIRECT

In some cases you will be cleared "direct" to the first VOR on the route; ATC means directly to the VOR in a straight line. They expect you to establish a course to the station and to stay on that course—keep the needle centered, the next VOR challenge.

To fly direct to a VOR, turn the OBS knob until the needle centers in the TO position and read the course in the window. That is the course to the station, and that is the course ATC expects you to fly.

You know whether the wind is from the left or right and roughly how strong it is. After establishing the course to the station, fly a trial correction, left or right, of 2°, 5°, or 10° to keep the needle centered.

REFERENCE HEADING

Note the heading. I call this a "reference heading" or "holding heading" because it is the heading that holds the airplane on the correct magnetic course with the needle centered. Make small corrections left and right of the reference heading and the needle should hover around the center.

The overall procedure for establishing a reference heading and adjusting it to keep the needle in the center is known as "tracking." In addition to using tracking to stay on course direct to a VOR, tracking will also maintain a course along a prescribed airway.

Keep in mind that the wind will rarely remain constant for any length of time. So once you are established on your assigned radial or bearing and have worked out your holding heading, you will still have to adjust it a few degrees for slight changes in wind direction and strength. This, too, is part of tracking.

BRACKETING

In the real world of IFR, you may not always be able to determine wind strength and direction, especially if you fly through a front or other rapidly changing weather conditions. Or you might become momentarily disoriented and unsure of what correction to make to stay on course and keep the needle centered.

In either case, "bracketing" will get you back on course quickly and, at the same time, show what the holding heading should be to maintain that course.

Bracketing is a series of smaller and smaller turns from one direction to another across the desired course. Bracketing can be done on both inbound and outbound legs. Start by turning the airplane to a heading that is the same as the desired magnetic course. Make the first turn 30° toward the needle. Hold the correction and make the needle move back to the center.

When the needle has returned to the center, cut that first 30° correction in half, and turn 15° toward the needle. Make it move back to the center again. When it has centered cut the correction in half and turn 7½° toward the needle. Again, make it move toward the center. You will quickly find a reference heading that will position the needle near the center and stop it from moving. Make minor adjustments left and right of that holding heading to keep the needle centered.

CHASING THE NEEDLE

In describing bracketing, you will note that I was careful with each turn to say "make the needle move." What I meant by this is maintain the correction until achieving the desired result—in this case, returning the needle to the center.

Because the human eye is always quick to pick up motion, there is a great temptation to fixate on a VOR needle as soon as it starts to move and to turn toward it—to "chase" it. If you chase the needle making larger and larger corrections it will be impossible to predict when the needle will stop or reverse direction; pretty soon you will be way off

Fig. 10-7. *En route correction to return to a VOR course.*

Depending upon how far off course you are, use a 10, 20 or 30 degree intercept angle to return to the desired track (course).

210°/030°

Full Scale Deflection Line

360°

030°

Point of Intercept

Desired Course

course. If you chase the needle when half-a-mile from the station, you might make a 45° correction when you are only 50 feet off course and blow the station passage.

So set in a correction, hold it until the needle moves to the center, then adjust the correction. Don't start taking the correction out as soon as the needle starts to move.

STATION PASSAGE

If you bracket and track the VOR properly, the adjustments to your holding heading will become smaller and smaller near the station. This is very important, because your needle gets extremely sensitive the closer you get to the station.

If you are still making large heading changes close to the station, the airplane will pass way off to one side or the other. If the VOR just missed is the final fix on a VOR approach, execute a missed approach and try again.

Make your greatest efforts several miles out to establish the reference heading for perfect station passage. When you approach within a quarter of a mile of a station (and this is slant range), you enter a zone of confusion where none of the VOR instrument indications will hold steady.

Maintain the reference heading through this zone of confusion and note the time and the TO-FROM indicator flips to FROM. Maintain the reference heading—or turn to a new outbound heading and hold that—until the instrument indications settle down.

Don't chase the needle! As you maintain the reference heading, or turn to a new one, analyze whether you are to the right or the left of your outbound course, and then set up a re-interception or a bracketing procedure to get back on course. Then make minor adjustments to keep the needle centered.

PRACTICE PATTERNS

Patterns A and B are excellent for practicing VOR interception and tracking. Start the patterns over a VOR station, using a VFR altitude (odd or even thousand feet plus 500 feet) to avoid IFR traffic that may also be using the same VOR station. Start the pattern at the station and plan for each straight leg to return over the station.

While developing basic attitude instrument flying skills you will also hone VOR interception and tracking skills. And, as we have noted before, these patterns contain all the elements in an instrument approach. So, when practiced with a VOR station, they also provide an introduction to VOR approaches.

After your instructor has introduced these patterns, practice with a safety pilot. As always, discontinue practice if problems begin to crop up.

11
Holding Patterns

MANY PEOPLE ARE SURPRISED WHEN I INTRODUCE VOR HOLDING PATTERNS AS EARLY as Flight Lesson 5. There are two reasons for this.

First, holding patterns around a VOR station provide excellent practice in VOR interception and tracking close to the station. The needle indications are very sensitive; holding pattern practice will quickly sharpen tracking skills.

Second, there seems to be something awe-inspiring and difficult about holding patterns. They are really quite simple when you go about them the right way. Nevertheless, many pilots are wary of holding patterns, so I tackle them early in the course so that students will feel comfortable with them later on.

My technique for teaching holding patterns works very well. Initially, students learn how to fly the pattern while correcting for wind drift. Only after they are comfortable with the racetrack pattern and have mastered wind correction techniques do I then teach pattern entries.

FIGURE 11-1 illustrates the basic elements of a standard holding pattern around a VOR. The standard pattern has right turns, the inbound leg is one minute long. Non-standard patterns have left turns. Above 14,000 feet the inbound leg is 1½ minutes.

Pick a nearby VOR and fly inbound on any convenient course with the OBS needle centered. At station passage you will be very busy for a few seconds running an important checklist.

Fig. 11-1. *Elements of a standard holding pattern.*

FIVE Ts: TIME, TURN, TWIST, THROTTLE, TALK

Do these at station passage:

1. Note your *time* of arrival at the fix and write the time down on the log or directly on the chart.
2. Start a 180° *turn* to the outbound course with a wind correction as necessary.
3. *Twist* the OBS knob to the set new inbound VOR course.
4. *Throttle*. Reduce power to conserve fuel while in the holding pattern; you are not going anywhere.
5. *Talk*. Report reaching the holding fix. If in radar contact you do not have to re-port entering the hold unless requested by ATC. Always ask the question, "Do I need to report to ATC now?" During an instrument approach the answer to that question will very often be "yes." (When in doubt, report.)

These are the five Ts of instrument flying—*Time, Turn, Twist, Throttle, Talk*. In oth-er words, aviate, navigate, and communicate. You will encounter the Five Ts over and over and, especially during instrument approaches. Use this checklist with every holding pattern and keep using it until it begins popping into your head automatically.

After running through the checklist, the next task is to determine when to start timing the outbound leg. This is easy. After station passage, and while you are making the turn to the outbound leg, the TO-FROM indicator will show FROM. When you are abeam the holding fix heading outbound, the indicator will change to the TO position. Start timing when the indicator changes to TO.

WIND CORRECTIONS

You know from VOR tracking practice that wind is almost always a factor. How do you correct for the wind in a holding pattern?

As you flew inbound to the VOR, you kept the needle centered. So you should have a pretty good idea of the wind correction for the inbound leg upon reaching the fix. The wind correction angle on the outbound leg will be *double* and *opposite* correction on the inbound leg.

If you held a wind correction angle of 4° into the wind on the inbound leg, you should hold a wind correction of 8° into the wind heading outbound. The reason for doubling the wind correction angle is to compensate for the effect of the wind on those two one-minute turns at the fix end and at the outbound end.

Even if you fly for precisely one minute on the outbound leg, you will probably find that the inbound leg is not exactly one minute. Again, this is because of wind. The correction is simple; adjust the time of the next *outbound* leg to compensate for the difference.

If the inbound leg is only 45 seconds, for example, add 15 seconds to the outbound leg and make it 1 minute and 15 seconds. If the inbound leg is 1 minute 30 seconds, subtract 30 seconds from the outbound leg. In the first case you are compensating for a tailwind on the inbound leg; in the second case you are correcting for a headwind on the inbound leg.

EN ROUTE HOLDING

The preceding was a description of the procedure used for establishing an en route holding pattern. If there is a delay while en route, ATC may simply hold you on present course and same altitude at a convenient fix.

If no holding pattern is shown on the en route chart, ATC will state:

1. What fix to use and where the holding pattern will be located in relation to the fix (north, south, southwest, etc.).
2. What radial to use.
3. Nonstandard instructions, such as left turns and length of legs in miles if DME is utilized.
4. An "expect further clearance time" (EFC), or the time to "expect approach clearance (EAC)" if held on a segment of an instrument approach. These "expect" times will be given either as a specific clock time, such as 2045, or they will be given in minutes. It is not uncommon to hear ATC say "expect further clearance to XYZ in ten minutes."

EFCs and EACs are important for two reasons. They tell when ATC expects to clear you to resume your flight. You need to know this to adjust the holding pattern to arrive back over the fix at the "expect further clearance" time.

EFCs and EACs also tell you when to depart a holding pattern in the event of lost communications. If ATC does not issue an EFC or an EAC, be sure to request it.

A typical holding clearance might be:

"Cessna three four five six Xray, hold southeast of the Huguenot VOR on the
    one four five degree radial, maintain five thousand, expect further clearance
    at one two one five."

Many holding patterns are already depicted on en route and approach procedure charts. If the holding pattern is shown, the clearance will be simpler. For example:

"Cessna three four five six Xray, hold as published northeast of Lake Henry VOR, maintain five thousand. Expect further clearance at one two one five."

In this case, ATC would expect you to find Lake Henry VOR and fly the depicted pattern northeast of the VOR.

ATC recognizes that holding patterns can be somewhat imprecise because of the wind, the skill of the pilot, and the different airspeeds of different types of aircraft. So they establish a buffer zone on the holding side that is at least double the amount of protected airspace around the pattern.

HOLDING PATTERN ENTRY

When comfortable with the patterns, you will have also automatically mastered one of three methods of entering holding patterns recommended by the FAA. You will have used the *direct* method every time you set up a pattern. The direct entry, simply stated, means that you fly to the station and make a turn directly to the outbound leg, which is exactly what you have been practicing. On IFR flights, you will use direct entries for most holds issued by ATC.

The other two ways to establish holding patterns are the *teardrop* and the *parallel* methods (FIG. 11-2).

To enter a standard holding pattern with the teardrop method, cross the holding fix and proceed outbound at an angle of 30° to the holding course for one minute, then turn right to intercept the holding course.

For the parallel method, cross the fix and fly outbound parallel to the holding course for one minute, turn left, fly direct to the fix, then turn right to the outbound course.

Before going into further detail, some basic points will make holding entry much easier.

Direct, teardrop, and parallel entry methods are recommendations only, not required. The FAA prefers use of these methods. If you get mixed up and use a direct entry when you should have used a teardrop entry, you won't get into trouble as long as you establish an accurate pattern quickly and efficiently and do not violate the protected airspace.

You should become proficient in the three recommended methods, however. They are the best ways of entering holding that have been developed so far. And the examiner on the instrument check ride will expect them.

The entry and first time around the holding pattern are free. ATC understands that it is difficult to make a perfect textbook entry and fly a perfect one-minute racetrack on the first circuit. They won't bother you if you are a little off course on the first pattern, provided you maintain the correct altitude. But they do expect better tracking by the second circuit.

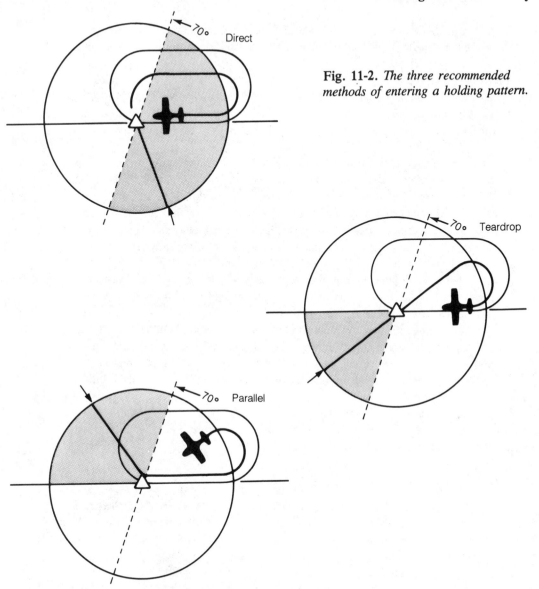

Fig. 11-2. *The three recommended methods of entering a holding pattern.*

CHOOSING THE CORRECT ENTRY

Turn again to FIG. 11-2. Note the line that has been drawn at an angle of 70° to the inbound leg of a standard right-hand pattern. (The pattern would be on the left side of the inbound leg for a nonstandard left-hand pattern, and the 70° line would be drawn in the opposite direction.)

The 70° line divides the area surrounding the holding fix into three "entry sectors," as indicated by the shading. The type of entry chosen depends on the entry sector in which you approach the holding fix.

I have flown with a few amazing people who can calculate entry sectors quickly in their heads down to the last degree. But this is the exception, not the rule. Most of us can't "do it by the numbers."

Fortunately, three other ways to plan holding entries have evolved over the years:

- Pencil in the holding pattern right on the chart (if it isn't already printed. Draw in an approximate 70° line—it's 20 degrees less than the perpendicular to the inbound course. Then draw the inbound course to the station. This will show which entry sector you occupy and which type of entry to choose. Remember, you don't have to draw the lines perfectly down to the last degree to set up an acceptable entry.
- Use one of the many plastic holding overlays available. Place the overlay over the holding fix on the chart. Align the holding pattern on the overlay with the inbound leg of the holding pattern. You will see clearly which entry sector you occupy. These clear plastic overlays usually have standard (right-hand) patterns on one side and nonstandard (left-hand) patterns on the other side.
- Use the heading indicator. Visualize the inbound leg of the holding pattern, then take a pencil or other straight object and rotate it 70° counterclockwise to this inbound leg. This will reveal the entry sectors very quickly. For a nonstandard (left-hand) pattern, turn the pencil clockwise 70° to indicate the entry sectors.

Finally, *visualization* is the key! Visualize the holding pattern in relation to the fix and where you are. Upon arrival at the fix simply make the shortest possible turn to join the race track of the holding pattern.

IMPORTANCE OF ALTITUDE CONTROL

ATC is more concerned about altitude control than making a perfect pattern entry because they might have planes above and/or below you in the same pattern. You might be in a pattern at 5,000 feet with planes below you at 3,000 and 4,000, and others above you at 6,000, 7,000, 8,000, and so on. This is a stack (FIG. 11-3). Stacks are a common occurrence around busy airports during landing delays due to weather.

In a stack, plans are cleared to proceed from the bottom, like dealing from the bottom of a deck of cards. Each time a plane is cleared to leave the stack, all the others are cleared sequentially to descend one level at a time. A vertical separation of 1,000 feet is maintained at all times. Obviously, there must not be confusion about altitude assignments, nor any sloppiness in maintaining assigned altitudes, with several planes all holding in the same racetrack pattern around the same fix, separated only by altitude. The goal of 2, 2, and 20 becomes a realistic requirement in this situation.

Fig. 11-3. *Stack, or shuttle descent in a holding pattern, for an instrument approach to an airport.*

difficulty maintaining a precise altitude. The goal of 2, 2, and 20 becomes a realistic requirement in this situation.

HOLDING PATTERN VARIATIONS

There are several variations in the basic en route holding pattern.

It is not unusual for an instrument flight to be kept in a holding pattern 20 minutes or more. When this happens it is perfectly all right to ask ATC for a pattern with two-, three-, or even five-minute legs. ATC will usually try to grant the request.

Longer legs make it easier to fly the pattern. More attention can be paid to altitude control and to establishing a precise inbound course with the needle centered. Longer legs are also much easier on passengers. Repetition of a straight leg for one minute, followed by a one-minute 180° turn could cause airsickness.

For planning purposes, a pattern with two-minute legs will take six minutes to complete, a pattern with three-minute legs will take eight minutes, and one with five-minute legs will take 12 minutes. If you expect to be in a holding pattern for any length of time, adjust power and re-lean the mixture to keep fuel consumption to a minimum.

INTERSECTION HOLDS

ATC will frequently issue en route holds at VOR intersections. Some are depicted on en route charts and when seen along the route, be mentally prepared to use it.

Intersection holds are easily managed. To reduce cockpit confusion, always set up the holding course on the top (No. 1) NAV. Set up the intersecting bearing on the No. 2 NAV.

If, on the No. 1 NAV, you always set up the course *to the station*, not the radial, the needle will always be located in the same direction as the VOR. If the needle is to the right, the station will also be to the right. This is more a matter of reducing cockpit confusion than anything else. Set up the two VORs as described above and you will always have a clear picture of exactly where you are in the pattern. (These are also good procedures even when no holding is involved and you want to keep track of the intersections along an en route leg.)

You have arrived at the intersection when the needle on the No. 2 NAV centers. Do the Five Ts checklist.

Start timing the outbound leg when the No. 2 NAV needle moves back on the same side as the VOR indicating that you have reached the abeam position. In short—with the No. 2 NAV set to the radial (from), if the needle and the VOR are on the same side, you have not arrived. If they are on opposite sides, you have passed the intersection.

ONE-VOR INTERSECTION HOLDS

It is possible to identify VOR intersections and fly holding patterns around them using only one VOR. Sounds difficult at first, I know, but many pilots are quite accomplished in using this procedure. For a long time in the early days of VOR, having even one VOR in the cockpit was considered a luxury, and dual VORs were almost unheard of. So one VOR did the work of two.

It's not a good idea to spend any time on one-VOR intersection holds at this point. If one VOR fails in flight, notify ATC immediately. Consider landing as soon as possible because if the other VOR should also fail, you would have a job on your hands getting down safely. ATC would have to do a lot of fast shuffling to reroute the traffic in your vicinity to ensure safe separation.

If you would like to try one-VOR intersection holds, have an instructor coach you through a few in the simulator.

Instructor Note. Here is the procedure for a one-VOR intersection hold. Track the inbound leg with the appropriate wind correction established and the needle centered. One minute before ETA at the intersection, reset the NAV to the frequency of the station providing the cross-bearing and reset the bearing to that of the second station.

On reaching the intersection, proceed normally. When established on the outbound leg, reset the NAV frequency and bearing back to the first station. At one minute outbound, turn again and get re-established on the inbound leg with the needle centered. Fly inbound for 30 seconds, then reset the frequency and bearing to anticipate the cross bearing.

Obviously, a lot depends on the student's ability to get established quickly on the inbound bearing with the needle centered. It's not a good idea to practice one-VOR intersection holds until the student has thoroughly mastered the art of working close to a VOR station.

DME HOLDING PATTERNS

Another holding pattern variation is based on DME distances, rather than time. If you have indicated on the flight plan that you have DME aboard (code letter A), you can expect to be issued a DME hold. DME holding uses the same entry and racetrack procedures except that distances (in nautical miles) are used in lieu of time. Both the distance of the fix from the navaid and the length of the outbound leg will be specified by ATC.

For example, if you are heading *toward* the VOR/DME (FIG. 11-4) with the fix distance 10 nm and the outbound leg 15 nm, you will enter the racetrack when the DME reads 10. You will end the outbound leg when it reads 15, and commence the turn back to the inbound leg.

Clearance for a DME hold would be something like this:

"Cessna three four five six Xray, hold 10 north of Carmel VOR on the three six zero degree radial, five mile legs, expect further clearance one five four five, maintain five thousand."

DME holding patterns can also be established with the inbound leg heading *away* from the VOR/DME, as shown in FIG. 11-5. In this example, the DME fix is 30 nm from the station and the end of the outbound leg is 20 nm from the station.

DME holding patterns are certainly a lot easier to manage, and this should be a factor when considering whether or not to invest in DME equipment. But don't throw away your stopwatch yet! You'll need it to time nonprecision approaches.

Fig. 11-4. *Holding pattern toward a DME facility.*

Fig. 11-5. *Holding pattern away from a DME facility.*

12
Stalls, Unusual Attitudes, and Partial Panel

THIS CHAPTER WILL COVER IFR STALLS, STEEP TURNS, CRITICAL ATTITUDE RECOVERY, and control of the airplane under partial panel conditions. Naturally, you will work on these situations and conditions under the hood.

Stalls under the hood?

Well, why not? Many people are surprised that my course includes stalls under the hood. However, it is very helpful for an instrument student to become familiar with the instrument indications and the feel of the airplane as its controllability degrades from minimum controllable airspeed into the power-off full stall.

In the real world of IFR you might find yourself in a power-off stall if you become distracted and reduce power too much or allow the airspeed to drop too low on an approach. Likewise, a power-on stall could develop at the missed approach point with the rapid application of power with gear and flaps down, especially if a considerable amount of nose-up trim has been set.

PRACTICING STALLS

You have been practicing flight at minimum controllable airspeed since Flight Lesson 2. Stall work will simply be an extension of minimum controllable airspeed.

In practicing the power-off stall, fly at minimum controllable airspeed with power set at 1,500 rpm or 15 inches of manifold pressure and flaps and gear down if applicable.

Instructor Note. Pick a convenient safe VFR altitude for practicing stalls and *always* do clearing turns. The student should do the clearing turns under the hood while setting up for minimum controllable airspeed. This will provide valuable additional practice in handling the airplane on instruments through a wide range of changing control forces. The instructor has sole responsibility for collision avoidance.

When straight and level with flaps and gear down and power at 1,500 rpm, reduce power to idle. Cross-check with the turn coordinator and add rudder pressure to keep the ball centered. Allow the airspeed to decrease while holding altitude constant until a full stall occurs.

STALL RECOVERY

As the airplane stalls, effect the stall recovery by applying full power. Reduce back pressure to reduce the angle of attack. Don't push forward on the yoke—you'll loose too much altitude if you do. Use a reduction of back pressure to pitch down slightly on the attitude indicator, then return to straight and level flight promptly without re-stalling.

Use the attitude indicator to maintain straight and level. Use rudder pressure, not ailerons, to hold heading while the airspeed is low.

In a stall, one wing will frequently drop. The reaction of many pilots in critical situations and at critical airspeeds is to use the ailerons to raise that low wing. This is incorrect in most airplanes and might make things worse. The proper procedure is to use opposite rudder to add a little speed to the slower, descending wing and give it lift.

Instructor Note. Because of deficient earlier VFR training, students might require extra practice in the use of rudder rather than ailerons to raise a low wing in a stall.

As the power becomes effective, start raising the flaps in increments. When you attain climb airspeed (V_y) in a straight and level attitude, gradually pitch up to the first line above the horizon on the attitude indicator. (In a retractable gear airplane, delay raising the landing gear until a positive rate of climb has been established.)

Retracting the Gear

Note that the flaps are raised before the gear is retracted. There are two reasons for this. First, the flaps create much more drag at low speeds than the gear. You may establish a climb faster by getting the flaps up first.

Second, when you execute the missed approach, you want to have a positive rate of climb before retracting the landing gear. If you have erroneously allowed the airplane to drift down, you will touch the runway with the wheels, not the first six inches of the propeller blades!

This stall maneuver combines both the power-off stall that might occur on the final of an instrument approach with the full-power, flaps-and-gear-down situation of a missed approach. The IFR stall maneuver should be practiced on instruments until you can quickly recognize the stalled condition when you begin to lose altitude, then effect a prompt recovery with minimal additional loss of altitude. After you become skillful at this, you should lose no altitude. Losing more than 25-50 feet in the recovery is unsatisfactory and could be hazardous to your health.

Recognize the stall quickly and execute the recovery procedures promptly, correctly, and automatically.

STEEP TURNS

The next maneuver was required to obtain a private pilot certificate—steep turns at 45° bank. The difference now is steep turns with the hood on, solely by instruments.

Instructor Note. The traditional method is to establish a 45° bank and then complete a 360° turn in one direction, followed immediately by a 360° turn in the opposite direction, rolling out on the original heading.

Common errors are altitude control—especially during and immediately following the roll out—and changing to the opposite direction. Students will usually lose altitude at 60° into the turn when executed to the left and gain altitude in a right turn because of "P" factor. During the change of direction there is also a tendency to gain altitude due to the excess back pressure required. An excellent training maneuver to conquer this problem if it persists is to have the student do a series of entries to steep (45° bank) turns in opposite directions until the problem is solved.

Steep turns require a faster scan to make sure you absorb all the information the instruments are showing when you need it. You can't afford to fixate on any single instrument because everything will be happening quickly. You will use the same control and support instruments for steep, 45° banked turns as standard rate turns, but move your eyes around the panel faster. Use the same scan pattern—attitude, heading, altitude—but do it faster.

One problem with this maneuver is that the inner ear senses a 90° bank with each change in direction. Beware of nausea; when air sickness occurs, all learning ceases.

Altitude Control

Cross-check with the altimeter to make sure you are maintaining altitude. On this maneuver you are allowed ± 100 feet. However, it is much easier to limit the variation in altitude to ± 20 feet than to allow the altitude to vary by 100 feet.

Continue the turn for a full 360°. Once established in this steep turn, of course, the bank tends to increase. If you are not paying attention or your scan is too slow, you will experience a rapid loss of altitude and an ever-steepening bank. Correct this by reducing bank to 20° or 30° to recover the lift lost in the steep turn. It's almost impossible to regain the lost altitude unless you decrease the angle of bank.

Trim

Students almost always ask what to do about trim in steep turns. I recommend that the maneuver be performed a few times without adjusting the trim to feel what the control pressures are like and what it takes to cope with the control pressures solely by instruments.

Add a little power—100 rpm or so—in the turn to help keep the nose up. Add a touch or two of nose-up trim. With practice the combination of added power and nose-up trim will result in smoother, easier steep turns.

After completing a 360° turn in one direction, make a smooth transition to a steep turn in the opposite direction. Do not pause during the reversal and do not fly straight and level between one turn and the other.

Keep the attitude indicator dot right on the horizon as you roll and do not allow altitude to vary more than ± 20 feet for an easier time managing the control pressures in the reversal. You have been applying so much back pressure (or adding power and nose-up trim) to maintain altitude during the 45° turn that when you roll into this reversal, it almost feels like you have to push the nose forward to keep from climbing. Also, lift increases when the wings roll through the level position. Fix that dot on the horizon line and visualize rolling around it from one direction to another.

COPING WITH VERTIGO

Steep turns under the hood might induce vertigo. In its milder forms, vertigo can make a pilot feel like he is in a climbing or diving turn when you are flying straight and level.

In extreme vertigo cases, it might feel like a straight-and-level descent when you are actually plummeting down in a tight, ever steepening turn. This is the infamous "graveyard spiral" that claimed so many early air mail pilots flying at night or in the clouds before the introduction of adequate flight instruments. The spiral also claims many VFR pilots who continue flying into IFR conditions.

Fortunately, with reliable instruments and better knowledge about human physiology, there is no reason to succumb to vertigo, or spatial disorientation as it is technically called. The one cardinal rule for coping with any kind of vertigo, dizziness, or confusion between what you see on your instruments and what your other senses are telling you is trust your instruments. Accept the fact that in many circumstances the senses are wrong when instruments are correct.

Here's what happens: When flying VFR eyes are the main source of information about motion and position, like eyes are primary sensors on the ground. Pilots control the attitude of an airplane by seeing that the nose is above, below, level with, or banked in relation to the horizon and other outside references. The mind subordinates the input of other senses—hearing, smell, motion—to the messages coming from the eyes.

When pilots fly on instruments, eyes are seeing symbolic rather than actual references. The input is not as strong nor as vivid as the real world. Other senses—particularly the motion senses—tend to take over the eyes' role.

Semicircular Canals

"Semicircular canals" located in the inner ear sense motion. The canals are three tiny tubes set roughly at right angles to each other. The tubes are filled with a fluid that is set in motion with movement. The fluid's direction is sensed by tiny hairs and transmitted to the brain.

But the fluid does not keep moving. Unless there is further acceleration or deceleration, the fluid slows down and motion ceases. The hairs return to their normal position. If you roll into a turn and hold it steady for 30 to 45 seconds, the fluid in the semicircular canals stops moving and the hairs cease sending the turning message.

If your senses are screaming "straight and level" while actually in a turn, you might be tempted to increase the angle of bank. This is how the ever-tightening turns of the graveyard spiral begin.

With more experience controlling the airplane solely by reference to instruments, the sense of sight will override the sense of motion. But vertigo is an insidious problem. It can creep up on even the most experienced instrument pilots if fatigue sets in. Illness, medication, alcohol, sleep loss, and mild hypoxia (lack of oxygen) can increase susceptibility to vertigo, as can extended climbs with the ball off center.

Returning now to the 45° banked-turn maneuver, start from straight and level at some convenient cardinal heading such as 360°. Roll into the turn and establish a 45° bank with the attitude indicator. Place the dot right on the horizon line and visualize the airplane nose rolling on that dot.

UNUSUAL ATTITUDES

A steep turn with a bank that increases beyond 45°, coupled with a rapid loss of altitude, opens the realm of unusual attitudes.

From the beginning, my students work on unusual attitudes with the attitude indicator covered up. It is advisable not to practice initially with a full panel. (This is the instructor's decision.)

Let's consider what happens when an attitude indicator begins to fail. The first point is that it takes time for the gyros to wind down. Even if there is a sudden failure of the vacuum system powering the gyros of the attitude indicator, the gyroscopes in the instrument will lose their momentum slowly.

You might not be aware that a failure has occurred. The attitude indicator doesn't suddenly roll over and die at a dramatic angle; the indicator gradually drifts off. You might continue to use the attitude indicator as the control instrument while it is gradually leading you astray.

The first assumption dealing with an unusual attitude is that the attitude indicator has failed. Don't stop to analyze the failure. Assume that it has occurred, deal with the unusual attitude immediately, and when everything is under control again, try to figure out what went wrong.

RECOVERY PROCEDURES

The first instrument to check in an unusual attitude is the airspeed indicator. Its indication will determine what actions to take.

If the airspeed is *increasing* the airplane is in a dive and might run out of altitude during an approach or departure, or exceed the red line airspeed. (V_{NE}, never exceed speed) if you are at altitude.

The recovery procedure for a *diving* unusual attitude is:

1. Reduce power to idle.
2. Level the wings with rudder and aileron. Center the needle of the turn coordinator to control the bank. Keep the ball centered.
3. Raise the nose to stop the descent. Refer to the airspeed indicator to increase pitch and stabilize the airspeed at cruise.
4. When you attain cruise airspeed, apply cruise power and establish straight and level flight on partial panel.

I will talk about how the instruments are used for partial panel flying in a few paragraphs.

Let's cope with an unusual attitude in which the airspeed is *decreasing*. The recovery from a *climbing* unusual attitude is:

1. Add full power to increase airspeed and reduce the risk of a stall.
2. Lower the nose. Don't run out of airspeed and get into a power-on stall. Lower the angle of attack. Use the airspeed indicator to decrease pitch and return to cruise airspeed.
3. Level the wings with rudder and aileron. Stabilize the turn coordinator to control the bank. Keep the ball centered.
4. When the airspeed reaches cruise, reduce to cruise power and establish straight and level flight by partial panel.

Summary of Unusual Attitude Recovery Procedures

Resumption of control is initiated by reference to airspeed, altimeter, VSI, and turn coordinator.

Nose Low, airspeed increasing:
1. Reduce power.
2. Level the wings.
3. Raise the nose.

Nose high, airspeed decreasing:
1. Increase power.
2. Lower the nose.
3. Level the wings.

Common errors:

1. Fixation. Staring at the least important instrument. (Keep eyes moving.)
2. Improper trimming or continuously holding pressure against the trim. (Learn proper trimming.)
3. Cockpit disorganization. (Plan to have everything in proper sequence and readily available.)
4. Attempting recovery by the seat of the pants, which produces misleading memory sensations. (Believe the instrument.)

Finally, as the famous violinist told the budding musician who inquired about how to get to Carnegie Hall: Practice! Practice! Practice! Practice!

DEGREE OF REALISM

Practice unusual attitudes in realistic circumstances. I have the student start a turn and lower his head to try to find an obscure intersection on an en route chart or review a missed approach and holding instructions. Trying to find something complicated while in a turn is certainly a realistic and common scenario in the real world of IFR.

Usually when I say "OK, you have the airplane!" the student snaps his head up so quickly that he also gets a touch of vertigo to add further realism to the scenario.

The maneuver need not be violent. A nose-high or nose-low 45° bank is usually sufficient to train the student to make the correct responses.

I would like to simulate failure of the attitude indicator by turning it off and letting it spin down. But because this is not possible I cover the attitude indicator with a round card cut to fit over the face. (They can easily be cut from file folders.) Affix a piece of electrician's tape to the top of the disk to hang the disk on the rim of the panel just above the instrument.

With this arrangement I can lift the disk up without removing it to get confirmation about how well the student is doing by comparing the partial panel with the attitude indicator. This confirmation works very well in building a student's confidence when he first begins to work without a full panel. On many panels there is even a little thumb-indent below each flight instrument that makes it easy to lift the disk up momentarily whenever the instructor wants to make a comparison.

Instructor Note. Covering instruments is also a good technique during full panel practice if a student habitually falls into the habit of fixating on one or two instruments. For example, if you observe the student paying more attention to the movements of the altimeter or VSI than to the attitude indicator, cover the altimeter and VSI occasionally to accustom the student to the function of the attitude indicator as the correct control instrument for pitch.

Failures of the heading indicator, likewise, are also very rarely dramatic. You might be flying along in fine shape and wonder about problems maintaining a VOR course while holding the reference heading perfectly. You reset the heading indicator to agree with the magnetic compass, but the problem doesn't go away.

It is hard to generalize about characteristics of a dying gyro in the heading indicator or in the attitude indicator or in both, which is the net result of having the vacuum pump fail. The best way to get an idea of how these two gyro instruments behave when they fail is to sit in the airplane for a few minutes after the engine is at the conclusion of a flight and watch the slow demise of the attitude indicator and heading indicator as the gyros wind down.

PARTIAL PANEL PROCEDURES

Recognition of instrument failure becomes the first step in partial panel work, regardless of whether the failure becomes apparent in an unusual attitude or in erroneous readings that gradually become more serious. Following recognition of the failure make a positive decision to disregard erroneous instruments and turn to other instruments to supply the missing information. (FIGS. 12-1, 12-2)

Some pilots find it helpful to cover a failed attitude indicator and heading indicator, and they carry little rubber suction cup soap dishes on every flight just for that purpose. This is a good idea. In a tight situation you don't want to inadvertently include failed instruments in the scan. If you don't have any soap dish suction cups handy, just tear off pieces of paper and use them to cover the failed instruments.

AIRSPEED DECREASING   GAINING ALTITUDE

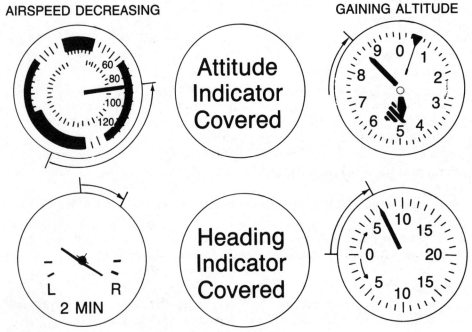

Fig. 12-1. *Nose-high unusual attitude on partial panel, with a steep right turn.*

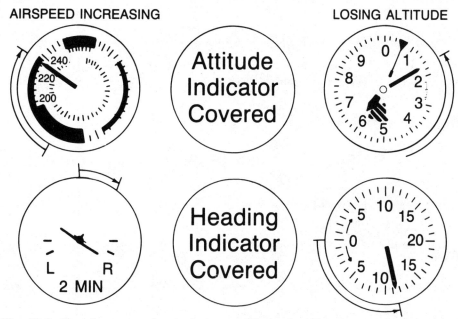

Fig. 12-2. *Nose-low unusual attitude on partial panel, with a steep right turn.*

The next step in the real world of IFR would be to land immediately after a major instrument failure. Don't hesitate to declare an emergency and get vectors from ATC to the nearest VFR airport. If no VFR field is available, head for the nearest IFR airport that has your personal minimums. There is absolutely no point in continuing an IFR flight with major instrument malfunctions. Things will only get worse as fatigue sets in.

PARTIAL PANEL CONTROL INSTRUMENTS

Because you can't always count on going VFR immediately after an instrument malfunction, be prepared to fly on partial panel. Other instruments replace the functions of the attitude indicator and heading indicator when no longer available (FIG. 12-3). Like the full panel (FIG. 8-5) a primary instrument provides the most pertinent information about how well you are doing and it does not move when flying precisely.

Use the same scan pattern for partial panel as full panel: attitude, heading, altitude. The only difference with partial panel is scanning two instruments for attitude information rather than one.

Note in FIG. 12-3 how the instruments are used in straight and level flight, for example. The control function, formerly performed by the attitude indicator, is now divided between the airspeed indicator and the turn coordinator.

Pitch is controlled with the airspeed indicator. If you are 100 feet below your desired altitude, use backpressure on the yoke to reduce airspeed five to seven knots. This will pitch the nose up slightly and regain the lost altitude. Conversely, if you are 100 feet above your desired altitude, you can increase airspeed by a very slight forward pressure on the

	CONTROL	PRIMARY	SUPPORT
STRAIGHT and LEVEL			
Pitch	Airspeed Indicator	Altimeter	VSI (Rate of Climb)
Bank	Turn Coordinator	Magnetic Compass	ADF
Power			RPM/MP
STANDARD RATE TURN			
Pitch	Airspeed	Altimeter	VSI
Bank	Turn Coordinator	Sweep Second Hand	ADF
Power		Airspeed	RPM/MP
CONSTANT AIRSPEED CLIMB			
Pitch	Airspeed	VSI	Altimeter
Bank	Turn Coordinator	Magnetic Compass	ADF
Power			RPM/MP
CONSTANT AIRSPEED DESCENT			
Pitch	Airspeed	VSI	Altimeter
Bank	Turn Coordinator	Magnetic Compass	
Power			RPM/MP
CONSTANT RATE DESCENT (ILS)			
Pitch	Airspeed	Glide Slope Needle	VSI
Bank	Turn Coordinator	Localizer Needle	ADF
Power		Airspeed	RPM/MP

CONTROL	=	Main reference instruments
PRIMARY	=	Key quality instrument
SUPPORT	=	Back-up or secondary instrument

*The Primary Instrument is always the instrument that gives the most pertinent information and is not moving when flying precisely.

Fig. 12-3. *Partial Panel Instruments: Control, primary, and support instruments for parial panel when loss of vacuum affects attitude indicator and heading indicator.*

yoke and pick up approximately five knots, indicating a slight nose-down pitch. This will gradually bring you back to the desired altitude.

When you reach the desired altitude and the airspeed stabilizes at the desired cruise speed, a minor trim adjustment might be required to help hold that altitude.

Keep wings level with the turn coordinator. As in full panel, keep both feet on the rudder pedals and apply aileron pressure if the coordinator leans steadily in one direction or the other. The turn indicator is very sensitive; don't attempt to make a correction every time it moves, or else you will begin chasing the needle and very quickly lose control of the heading. As always, keep the ball centered with rudder pressure.

The Poor Man's Heading Indicator

Here is an item I bet you won't find in any other book. Note that the ADF is listed as a support instrument for bank in straight and level and several other flight conditions. Why is this?

Because the ADF is an excellent source of bank information. An ADF is tuned to a station in front of the airplane, the ADF will very quickly indicate drifting left or right, the same way the heading indicator would if it were functioning. I call the ADF the poor man's heading indicator.

The ADF can also help you make standard rate turns. For example, if you have the ADF tuned to a station ahead and you want to make a 30° turn to the right, simply turn and begin the rollout approximately five degrees before the ADF needle has moved 30° to the left. There's more to the ADF than meets the eye!

With practice you will automatically look to the correct instruments for control references to replace the attitude indicator. As you make corrections and enter climbs, descents and turns, be careful to make minimal control movements. The greatest problem I find with instrument students flying partial panel is overcontrolling. Know exactly where to look for control references and minimize control movements and you will rapidly build up skills in the fine art of partial panel flying.

MAGNETIC COMPASS TURNS

Return to FIG. 12-3 and note how the magnetic compass starts to come into play as a primary instrument. The magnetic compass is one of the most familiar and perhaps least understood instruments in the cockpit.

It is extremely reliable, even though it bounces around a lot, and it is the only source of heading information that operates completely independent of all electrical, vacuum, and pitot-static systems. Pilots who understand the behavior of the magnetic compass and make a point of practicing with it frequently can achieve amazing flying precision with this instrument.

However, the magnetic compass "lies." You must know when and why.

VFR training taught how to use magnetic compass lag and lead to roll out on headings accurately. Here is a review:

1. When on a northerly heading and you start a turn to the east, the magnetic compass will initially show a turn to the west and will gradually catch up as the turn progresses to give an accurate indication when passing through 090°. Conversely, when on a northerly heading and you start to turn to the west, the opposite occurs.

2. When on a southerly heading and you start a turn toward the east, the magnetic compass will initially indicate a turn to the east, but will exaggerate the turn, gradually reducing the error and will be accurate when passing through 090°. When turning to the west, the compass behaves similarly by exaggerating the amount of turn initially, but in the correct direction. This lagging and leading behavior of the magnetic compass in 1. and 2. is "dip error" caused by the magnetic lines of force and their action on the magnetic compass when it is not precisely level. A rough formula for dip error is that it is equal to the closest latitude in degrees.

3. If you are on an easterly or westerly heading, Acceleration results in a Northerly turn indication while Deceleration results in a Southerly turn indication or error from your stabilized heading.

Remember this by the acronym ANDS:
   Accelerate
   North
   Decelerate
   South

Magnetic Compass Procedures and Exercises

1. When turning to a heading of south, maintain the turn until the compass passes south the number of degrees of dip error (your degrees of latitude) minus the normal rollout lead (one-half the angle of bank).

2. When turning to a heading of north, lead the compass by the amount of dip error (latitude) minus the normal rollout lead.

3. When turning to a heading of east or west, anticipate the roll out by the normal method.

The main point to remember is that on southerly headings, the magnetic compass precedes or leads the actual turn. On northerly headings, the magnetic compass lags behind the turn. FIGURE 12-4 shows the procedures for making magnetic compass turns in graphic form.

Turns to a heading based on indications from the magnetic compass are imprecise at best. However, you should practice magnetic compass turns on instruments until you can roll out on a specified heading ± 10° using the magnetic compass alone. But don't waste time trying to go beyond this. Concentrate on learning and recognizing the errors affecting the magnetic compass and never believe it unless you are straight and level and in stabilized flight.

This is especially important during the departure and approach phases of an instrument flight when the heading indicator must be accurately set. Do not touch the heading indicator unless you are in stable and straight and level flight. In a normal approach, the last opportunity for determining heading indicator accuracy might be the outbound (one-minute) leg of the procedure turn.

Fig. 12-4. *Magnetic compass leads (to the south) and lags (to the north) assuming a latitude of 30 for the dip error, and one-half a bank angle of 10 for leading the roll in, with acceleration/deceleration errors on east and west heading.*

TIMED TURNS

Partial panel heading changes don't always work so neatly that you can do them with the magnetic compass alone. Suppose you were on a heading of 035° and you needed to turn to 155°. How would you compute the rollout using the magnetic compass?

In this case you might be better off to use a standard rate turn timed with the sweep-second hand of the clock. Turn 120 degrees to get from your present heading of 035° to the new heading, 155°. At three degrees per second for a standard rate turn, you would execute a timed turn for 40 seconds.

Plan the rollout so that wings are level when time is up. If the rollout begins five seconds before the 40 seconds is up the heading change is exactly 120°. Make minor adjustments to the new heading as necessary after the magnetic compass has settled down.

For small changes of heading, use a half standard rate turn (one-and-one-half degrees per second). Roll into the half standard rate turn and count "one one thousand, two one thousand, three one thousand" up to the number of degrees of heading change you wish to accomplish. This will also bring you close enough to make minor adjustments after the magnetic compass settles down.

To summarize, practice steep turns, unusual attitudes, partial panel, and magnetic compass turns from Flight Lesson 6 through the end of the program. The new test standards require demonstration of these techniques during the instrument flight test.

13
The NDB Unmasked

THE NON-DIRECTIONAL BEACON (NDB) IS FOR EVERYONE. I LOVE IT, YOU SHOULD TOO. Not too long ago many people thought the NDB was on its way out, a relic of the past as far as modern IFR flying was concerned. But this has not proven to be the case.

NDBs have traditionally made it possible for small airports to have inexpensive and reliable instrument approaches when ILS and VOR installations are not feasible or too costly. For example, when I was chief flight instructor at Lincoln Park, New Jersey, I was able to get FAA approval for Lincoln Park's first instrument approach. It was an NDB approach based on the compass locator for the Morristown ILS 23 approach.

In addition to serving small airports, NDBs serve many other crucial functions at larger airports: fixes for holding patterns, procedure turns, missed approaches, and as compass locators for ILS approaches. Of the 170 instrument approach procedures in New York, for example, an astonishing 77 use NDBs either as the primary approach fix or as an essential component in other types of approaches.

The instrument flight test standards confirmed the continuing importance of the NDB by making mandatory an NDB approach be demonstrated successfully on the instrument flight test.

For some reason NDBs and their airborne ADF receivers have been a mystery for many years. The first aviation direction finding equipment was very complicated, no doubt about it. You had to turn a wheel in the ceiling of the cockpit that rotated a loop antenna attached to the fuselage. In photographs of older planes—as well on the older airplanes in museums—this loop is quite prominent.

That was not an ADF loop, it was a DF loop. There was nothing automatic about it. It had a direction finding antenna, but the pilot had to turn it manually, listen to the signal build and fade, then interpret the signal to determine the direction to or from the station.

The introduction of the automatic direction finding system or ADF, was a great advance. Now the indicator of the direction finding instrument automatically pointed to the station, no matter where the airplane was, and the ADF indicator was on the instrument panel, not over head or in back.

ADF ORIENTATION

This leads to the first concept to be stressed in ADF work: The needle always points to the station.

This is an obvious point, but many people do not clearly understand its implications. First of all it means that when tuned to an ADF station for homing, tracking, intersections, holding, or an approach, you never have to touch the system. Unlike the VOR, there's no OBS to think about, no need to "twist" anything at station passage or at any other time. This makes ADF much simpler to use.

Second, with ADF you always know where you are in relation to the station. There is no TO-FROM to interpret, no confusion about radials and bearings, no way to set the wrong OBS numbers.

ADF orientation is much simpler than VOR orientation. The head of the ADF needle with the arrow *always* indicates the *relative bearing* to the station.

To determine *magnetic bearing* to the station, simply add magnetic heading and relative bearing. In other words, magnetic heading plus relative bearing equals magnetic bearing to the station. You undoubtedly learned this in VFR training, but let's do a quick review now for some hints to simplify the process.

If you are on a magnetic heading of 030° and the ADF needle is 90° to the right, the magnetic bearing to the station is 120°.

$$\begin{array}{r} 030 \\ +090 \\ \hline 120 \end{array}$$

If you turn to 120°, the ADF needle will point straight ahead.

You don't even need to make a turn to confirm this. Take a medium length pencil and place it on the needle of the ADF, much in the same manner as a parallel ruler. Move it onto the heading indicator and the pencil will point to the magnetic bearing to the station, eliminating the arithmetic.

An inexpensive feature on many ADF indicators is a third, even simpler method of determining magnetic bearing to the station: the rotating azimuth ring. Simply rotate this ring manually to line up the magnetic heading with the mark at the top of the ADF indicator and the ADF needle will automatically point to the magnetic bearing to the station.

Let's say you are on a heading of 300° and the needle is pointing to the right wing of the airplane at 090°. Apply the formula and add 300 plus 090 and come up with 390. If the answer is more than 360°, all you have to do is subtract 360° from the total for the magnetic bearing to the station.

$$
\begin{array}{r}
300 \\
+090 \\
\hline
390 \\
-360 \\
\hline
030
\end{array}
$$

Instructor Note. Check the ADF visualizing devices and teaching aids available through aviation supply companies. These simple, inexpensive plastic or cardboard devices are invaluable for demonstrating ADF problems on the ground. Students can practice with them at home. This will save time and money over trying the same exercises in the air. And it will make your job easier.

Be careful, as in VOR, to tune the station correctly and verify its morse code identifier. Turn up the audio sufficiently to hear the ADF signal faintly in the background—but not so loud to interfere with communications—and keep the volume at that level.

When there is a disruption in VOR signals, warning flags appear on the face of the NAV instrument; this does not happen on the ADF indicator. Continuously monitor the identifier to detect any signal disruptions. Keep the volume low enough to hear the signal in the background. This is the only way to be sure that the ground station has not gone off the air for some reason or that the ADF unit in the airplane has not malfunctioned.

Check the heading indicator against the magnetic compass at least every 15 minutes and reset it as necessary throughout all phases of both IFR and VFR flight. But it is critical in ADF work because it is impossible to determine the magnetic bearing if the heading indicator has drifted off. Prior to a magnetic bearing determination, verify accuracy of the heading indicator.

ADF TIME/DISTANCE CHECKS

Another exercise to develop NDB orientation awareness is the ADF Time/Distance Check. Start in a simulator, which often reinforces the ability to orient around an NDB so well that a little additional practice in flight is needed.

Instructor Note. For airborne ADF practice of the Time/Distance Check and other ADF maneuvers, use commercial broadcast stations that lie well outside airways, approach and departure corridors, and traffic patterns. Very few are shown on Sectional Charts.

Look them up in a publication called *Flight Guide* sold by many FBOs and aviation supply houses. It comes in Eastern and Western U.S. editions. For further information contact:

Airguide Publications
1207 Pine Avenue
P.O. Box 288
Long Beach, CA 90801

Pick stations with the highest output for the best results. The needle might wobble with a commercial station. This usually clears up closer to the station and it is not a problem with powerful stations.

Tune the station and determine the relative bearing. Turn the number of degrees necessary to place the ADF needle at either 090° or 270° relative—the right wing or left wing.

Note the time and fly a constant heading until the bearing changes 10°. Note the number of seconds it takes for the bearing to change 10° divide by 10 for the time to the station in minutes.

$$\frac{\text{Time in seconds}}{\text{Degrees of bearing change}} = \text{Minutes to station}$$

You can determine distance to the station with this formula:

$$\frac{\text{TAS} \times \text{minutes flown}}{\text{Degrees of bearing change}} = \text{Distance to station}$$

INTERCEPTING A BEARING

Let's take an example from the real world of IFR. Suppose you have executed a missed approach and want to return to an NDB via a specified bearing to be in a good position to make another approach.

First, turn to the desired bearing.

Second, note the number of degrees of needle deflection left or right and *double* this amount to determine the interception angle.

Third, turn toward the needle the number of degrees determined for the intercept face.

As you do so the needle will appear to move away from the zero position on the ADF indicator face.

Wait a minute, the needle always points to the station. Why does it appear to move? The answer is that the needle doesn't move, the airplane moves and the ADF indicator face is attached to the airplane. In a turn, the needle continues to point to the station, but the airplane is moving under the needle.

A good, simple way to visualize this is to place a book or other object on the floor to simulate an NDB station. Stand a few feet away from the object, and point toward it so your arm simulates the ADF needle. Your nose becomes the zero point on the indicator face, matching the nose of the airplane.

Now turn your body to a new "heading" while continuing to point toward the "station." Your arm will behave the same way as the ADF needle, apparently moving away from your nose. But you will quickly see that it is really your body that is turning while your arm continues to point steadily at the "station."

Back to intercepting the bearing.

The fourth step is to maintain the new interception heading until the needle is deflected on the opposite side of the nose the same number of degrees as the interception angle.

Then turn to the desired bearing.

See FIG. 13-1. Let's say you want to head inbound to the station on the 360° bearing. Turn first to a heading of 360° and note a relative bearing of 15° to the left (A).

Next, *double* this relative bearing of 15° for an intercept angle of 30°. You turn toward the needle 30° (B). Maintain this intercept heading until the ADF needle has deflected 30° beyond the zero position (C).

Finally, turn inbound on a heading of 360°, the same as the inbound bearing to the station (D). (Lead the turn by 5° to avoid overshooting.)

With no wind, all you would have to do is continue the inbound heading to the station. But there is *always* wind. So let's proceed to examine the techniques used to correct for the effects of the wind.

HOMING IS UNACCEPTABLE

Heading inbound you can "home" in to the NDB by placing the ADF needle on the nose and keeping it there with heading adjustments. With the wind constantly pushing from one side, you will have to constantly change the heading as you proceed toward the station to keep the needle on the nose.

See FIG. 13-2 for an illustration of the homing method of reaching an NDB. The airplane starts heading inbound on the 360° bearing at A. As the wind blows from the left, the heading has to be adjusted to maintain the needle on the nose (B and C). At D, the airplane has been blown so far off the inbound bearing that it is flying a heading of 315° instead of the desired 360° to keep the needle on the nose.

Homing is unacceptable for IFR navigation because you stray too far from the intended course. The wide, looping course shown in FIG. 13-2 could take you into a hill, radio tower, or other obstruction at the minimum altitudes of an NDB approach. And FIG. 13-2 is not an exaggeration. Many poorly prepared instrument students do this on the instrument flight test.

TRACKING AND BRACKETING

Once you have intercepted the desired bearing, hold that heading and see what affect the wind has. Let's continue with the previous example—interception and tracking of the 360 bearing to the station—and see what happens.

Inbound Bearing

Fig. 13-1. *Intercepting an ADF bearing.*

Fig. 13-2. *The problem with ADF homing is that it takes you off course to a degree that is unacceptable for IFR flying—especially on NDB approaches.*

This is illustrated in FIG. 13-3. As the airplane proceeds toward the station, the wind from the left blows it off the bearing. At B the airplane has drifted off the bearing by 15° relative. To get back on course, double the drift noted at point B and turn toward the needle that amount, in this case 30° (C). You are reintercepting the bearing.

The ADF needle will then swing over to a relative bearing of 015°. Hold the intercept heading (330°) until the relative bearing reads 030°. You are back on the desired bearing to the station (D). Now reduce the corrections by half—15° in this case—to compensate for the wind (E). (You should also lead the turn back by approximately five degrees.)

This method of correcting for wind drift is called bracketing and you may have to "bracket" several times to establish a reference heading that will remain on the desired bearing, especially if a long distance from the station. The initial wind correction might be too large or too small to stay on the bearing. If so, adjust the correction. In the example above, for instance, if a 15° correction proves to be too much, reintercept the bearing and try a 10° correction.

Chasing the needle is a common mistake in ADF intercepts, tracking, and bracketing, just like VOR navigation. It is so tempting to follow that moving needle! Resist the temptation. Hold the heading steady until the needle reaches the relative bearing you want, then make the turn.

Near the station the ADF needle will become "nervous," start oscillating, and become more sensitive. Don't chase the needle, just fly the reference heading, or favor it by five degrees.

Passing over the station the needle will commence a definite swing to the right or left. Note the time and start timing the outbound leg.

Wait until the needle has definitely swung around to verify station passage—at least five to 10 seconds past the ADF station. At this point you should turn to the outbound magnetic course to determine on which side of the bearing you are now situated and how much you will have to correct.

OUTBOUND BEARINGS FROM THE NDB

The procedures for intercepting, tracking, and bracketing outbound from the station are almost identical to those used on the inbound magnetic bearing.

Turn to the outbound bearing and determine how many degrees you are off the bearing. Then double the error and turn toward the desired bearing by this amount. If you are off the bearing by 10°, turn toward it by 20°, for example.

When the angle of the ADF needle off the tail and the intercept angle are the same, you are on the desired bearing. Turn toward the outbound bearing and bracket outbound to determine the wind correction necessary to hold that bearing.

Remember that the needle always points to the station. Never put the needle on the tail by changing the heading. This will cause you to miss the bearing completely and, as you will see during an NDB approach, if you lose that outbound bearing, you will miss the airport.

Fig. 13-3. *ADF tracking corrections utilizing the bracketing procedure.*

Try to visualize where you are at all times. If in doubt, turn to the outbound bearing and check whether you are to the left or right of course, and by how much. Then double this amount and reintercept the outbound bearing. Practice this at altitude until you become thoroughly proficient at tracking outbound before commencing NDB approaches.

Better yet, if you are having a problem tracking NDB bearings outbound, try the visualization exercise again. Place an object on the floor to represent the NDB, and walk through the entire procedure of tracking a bearing inbound, then station passage, then tracking the same bearing outbound, using your right hand to always point to the NDB and your nose as the nose of the airplane. Believe me, this works!

I am very serious about practicing this visualization exercise until you understand clearly what the ADF needle is indicating especially outbound from the NDB. This is the critical final approach leg on an NDB approach. One of the most frequent causes for failures on flight tests is confusion on the final leg of the approach to the airport. Time and again a candidate will track inbound to the NDB accurately, then turn the wrong way after passing the station.

In so many elements of IFR flight, visualization is the key to making the right moves. The visualization exercise just described doesn't cost a cent, no matter how often you practice it.

PRACTICE PATTERNS

A good exercise for sharpening your skill at tracking and bracketing both inbound and outbound is the simple pattern depicted in FIG. 13-4. It is a good idea to use a commercial broadcast station for practicing this and other patterns and avoid straying into busy airspace.

Pick a cardinal heading such as 270°, and intercept the bearing inbound to the station as shown. After station passage, track outbound, correcting for wind for three or four minutes.

Fig. 13-4. *Pattern for inbound-outbound ADF tracking practice, with 90-270 course reversals.*

Then reverse course with a 90-270. Make a standard rate turn in one direction for 90° then reverse smoothly into a standard rate 270°, turn in the other direction. The 90-270 is quick, efficient 180° change in heading. And if you have maintained a steady bearing outbound, the 90-270 will place you close enough to the inbound bearing so that only small corrections will be required.

Track inbound to the station, then outbound on the other side for the same amount of time. Do another 90-270, and repeat the process until you can maintain steady bearings with corrections. A standard procedure turn may be substituted for the 90-270 course reversal.

Other good ADF exercises are Pattern "A" and Pattern "B" set up on NDB or commercial broadcast stations. Start each pattern over the station and orient the pattern on cardinal headings, at least in the beginning. Plan for each straight leg to return over the station.

These two patterns will provide plenty of practice in intercepting bearings, and tracking inbound and outbound. And, because the patterns contain all elements of an instrument approach, they are good introductions to NDB approaches, which will be discussed in the next chapter. "A" and "B" are also good exercises to practice in a simulator.

ADF HOLDING PATTERNS

Nothing could be simpler or easier than an NDB holding pattern. Track inbound on the desired bearing, making wind corrections as needed to maintain the bearing. At station passage, begin a standard rate turn to the right (left in a nonstandard holding pattern). Roll out on the reciprocal of the inbound bearing and double the wind correction to account for the effect of the wind on the outbound leg and in the two turns.

Mark station passage when the ADF needle reaches the 90° position. Adjust the timing of the outbound leg (nonprecision side) to produce a one-minute inbound leg. Now here's a good cross-check that many instructors overlook. After flying outbound one minute, the needle should point 30° off the tail—30° off to the right in a standard holding pattern, 30° off to the left in a nonstandard pattern. On completing the outbound leg, turn again, rolling out on the inbound heading. Note the bearing error and correct for it.

Once again, the quickest way to visualize ADF holding patterns is to run through the floor exercise described earlier in this chapter.

Review VOR holding for methods of entering holding. The same procedures apply to ADF holding.

Instructor Note. Students will get the picture on ADF holding patterns much quicker if you start them with direct entries, then progress to teardrop and parallel entries. Note also that holding patterns are excellent exercises is a student has difficulty with ADF procedures close to the station. You might extend the legs of the pattern to two or three minutes to provide more practice in bracketing and tracking. If you are in an ATC environment, get permission to extend the legs.

14
Approaches I: NDB, VOR, and DME

O NE OF THE MOST SATISFYING THINGS YOU AS AN INSTRUMENT PILOT WILL EXPERIENCE is breaking out at minimums on an IFR approach with the runway straight ahead and a comfortable landing assured. The average VFR pilot—and most passengers—consider this nothing short of miraculous!

But it isn't, really. By the time you begin to polish the fine points of IFR approaches, you will have learned to use the ADF and VOR/LOC equipment with great precision. And, if your instructor has been following the syllabus, you will have made several unhooded instrument approaches—enough for the "big picture" of what an instrument approach involves.

This chapter will cover the most common types of "nonprecision" approaches. I will show how to analyze an approach while planning for an IFR flight, and move step by step through representative examples of ADF, VOR, and DME approaches.

Understand, however, that every approach is different. The examples will illustrate the usual sequence of events in nonprecision approaches, along with the dialogue with approach control that accompanies this representative sequence. You will be able to apply the approach procedures, techniques, and communications to all nonprecision approaches. But always remember that approaches in the real world of IFR will differ from these representative examples. Adjust your procedures and communications accordingly.

NONPRECISION APPROACHES

A nonprecision approach is defined in the AIM Glossary as "a standard instrument approach procedure in which no electronic glide slope is provided." ADF, VOR, DME, and several less common types of approaches fall into the nonprecision category because they do not provide electronic glide slopes that give precise altitude information. The only electronic guidance they provide is for the approach course or azimuth information.

PRECISION APPROACH

A precision approach, on the other hand, is defined as "a standard instrument approach procedure in which an electronic glideslope/glidepath is provided."An ILS provides an electronic glideslope and is thus a precision approach. PAR (precision approach radar) depicts on the radarscope an electronic glidepath along which the airplane is guided by the final approach controller. In the military, this is known as GCA, or "ground controlled approach."

I will discuss precision approaches in the next chapter. However, nonprecision and precision approaches have many elements in common.

Common Elements

First and foremost, all instrument approaches have an altitude below which you cannot legally descend unless you are in a position from which a safe landing can be made. For nonprecision approaches this is called the MDA, or Minimum Descent Altitude. Precision approaches call this the DH, or Decision Height. Other common elements include an Initial Approach Fix (IAF), a Final Approach Fix (FAF), a final approach course, a missed approach procedure, and very often one or more intermediate fixes between the IAF and the FAF.

Because the term "segment" will be used frequently, here is what the different segments mean:

Initial Approach Segment. The segment between the IAF and an intermediate fix, *or* between the IAF and the point where the airplane is established on an intermediate course or the final approach course.

Intermediate Approach Segment. The segment between the IAF and the FAF.

Final Approach Segment. The segment between the FAF and MAP (Missed Approach Point).

Missed Approach Segment. The segment between the MAP, or arrival at the DH and the missed approach holding fix.

ALTITUDE MINIMUMS

Let's explore the question of how altitude minimums are derived for nonprecision approaches. The most important consideration, for obvious reasons, is safe obstacle clearance. This is spelled out in the TERPS, *United States Standard for Terminal Instrument Procedures*, the "bible" on instrument approach tolerance published by the FAA.

Along the centerline of the approach course, minimum obstacle clearance is provided for nonprecision approaches as follows:

- NDB located on airport, 350 feet.
- NDB off airport with FAF, 300.
- VOR located on airport, 300 feet.
- VOR off airport, 250 feet.
- DME arc as final approach course, 500 feet.
- Localizer, 250 feet.
- ASR radar (no glidepath), 250 feet.
- DF steer approach, 500 feet.

Obstacle clearance is provided for a "primary" area on either side of the final approach course centerline. The width of the primary area varies with the type of approach and distance from the field. But it is never less than one mile on either side of the final approach course centerline of an NDB, VOR, or other nonprecision approach.

These are comfortable obstacle clearances for the precise pilot, but there is not much room for error or sloppy procedures. Coming in from the FAF on an NDB approach, for example, you will clear obstacles a mile on either side of the inbound course by only 350 feet.

One of the most frequent reasons for failing the instrument flight test is going below minimums on an approach. I don't mean just momentarily dipping below a minimum because of turbulence, then correcting right away. What always surprises me are the candidates for an instrument rating who consistently fly 25, 50, or even 100 feet below minimums without taking corrective action, or who have not determined the correct minimums to begin with.

Minimums are so basic, yet many pilots seem to have problems with them. Why is this so? I believe it is because pilots do not always use a systematic procedure to analyze the minimums they will have to work with on a given approach.

ADJUSTMENTS TO MDA

Let's make a step-by-step analysis of a conventional nonprecision approach. If you follow these steps every time you plan a flight, you will develop the good habits that will enable you to quickly size up unfamiliar approaches that might be thrown at you at the last minute because of a change of runway.

The example for this exercise is the NDB RWY 36 at Oxford-Waterbury, Connecticut, an uncontrolled field near Westchester County Airport, New York (FIG. 14-1).

Let's analyze the altitude minimums at Oxford in a systematic way.

1. *Read the fine print first.* Don't leave the fine print until last because you might miss something very important. Consider items A and B on the Oxford NDB 36 approach chart:

Fig. 14-1. *A typical NDB approach, the NDB RWY 36 approach at Oxford/Waterbury-Oxford, Connecticut.*

A. "748' electric transmission towers running NE to SW 1,000 feet from approach end of Rwy 36."
B. "Obtain local altimeter setting on CTAF; when not available, use White Plains altimeter setting and increase all MDAs 160 feet."

You certainly need to know about significant obstructions in the approach path! And you will need an accurate altimeter setting for an instrument approach.

Here are some gems from other approaches:

"Mountainous terrain higher than airport in all quadrants." (Altoona-Blair County, Pennsylvania.)
"Updrafts may be encountered in vicinity of power plant 1.5 miles NE of runway threshold." (Islip/Long Island MacArthur, New York.)

And this sobering note from Indian Mountain Air Force Station, Alaska, as reported by Barry Schiff in *AOPA Pilot:* "CAUTION: Rwy located on slope of 3,415' mountain . . . successful go around improbable."

Yes indeed, the fine print can be *very* important. I suggest that you take a few minutes to browse through the approach chart booklet for your area and highlight the fine print at any airport you are likely to use. There may well be some surprises. But if you mark them, they won't surprise you during an approach.

Height of Obstacles

2. *Check the height of obstacles in the vicinity of the airport.* These obstacles determine the MDA. Look at the airport diagram for the Oxford NDB 36 approach (C). Note the number of obstacles that rise above 700 feet near the runways. Add 300 feet for the standard obstacle clearance for this type of NDB approach, and you will see that the published minimum altitudes for an NDB approach will be almost 1,100 feet.

Airplane Approach Category

3. *Pick the published minimum for your airplane category and type of approach,* either straight-in or circling (D).

Aircraft approach categories are explained on the inside front cover of the NOS approach chart booklet. The explanation is clear and simple:

"Speeds are based on 1.3 times the stall speed in the landing configuration at maximum gross landing weight. An aircraft shall fit in only one category. If it is necessary to maneuver in excess of the upper limit of a speed range of a category, the minimums for the next higher category should be used. For example, an aircraft which falls in Category A, but is circling to land at a speed in excess of 91 knots, should use the approach Category B minimums when circling to land. See following category limits:"

MANEUVERING TABLE

Approach Category	A	B	C	D	E
Speed (Knots)	0–90	91–120	121–140	141–165	Abv 165

Most general aviation propeller-driven airplanes fall in either Categories A or B. Categories A and B are the same at Oxford, but this is not always the case.

Straight-in vs. Circling

MDAs for each category are further classified by the type of approach, either straight-in ("S-36" at Oxford) or circling.

Straight-in approaches are allowed when the angle of convergence between the final approach course and the extended runway centerline does not exceed 30°. If the angle is greater than 30°, you must use circling minimums. Note that the pilot does not make the decision as to whether an approach is straight-in or not. Yes, you may break off a straight-in approach and circle to land on another runway (using the higher minimums), but the designation of an approach as straight-in or circling is based on the layout of the airport, the angle between the final approach course and the landing runway, the location of the electronic facilities, and the design of the instrument approach.

With these points in mind you can establish the basic minimums for each category airplane for the NDB 36 approach to Oxford:

S-36: 1,180-1 459 (500-1)
Circling: 1,300-1 573 (600-1)

The numbers mean:

- 1,180-1 and 1,300-1 are MDAs and the minimum visibility, 1 mile, for the two types of approaches.
- 459 and 573 are the heights above the airport (HAA) at the MDA. (These minimums for Oxford are valid *only* if you have obtained a *local* altimeter setting. (See the fine print.)
- (500-1) and (600-1) are military minimum ceilings and visibilities and are not applicable to civilian aircraft.

The MDA and visibility minimums for the NDB 36 approach at Oxford are 1,180-1 for straight-in, 1,300-1 for circling.

Instrument Approach Procedures (Charts)
INOPERATIVE COMPONENTS OR VISUAL AIDS TABLE

Landing minimums published on instrument approach procedure charts are based upon full operation of all components and visual aids associated with the particular instrument approach chart being used. Higher minimums are required with inoperative components or visual aids as indicated below. If more than one component is inoperative, each minimum is raised to the highest minimum required by any single component that is inoperative. ILS glide slope inoperative minimums are published on instrument approach charts as localizer minimums. This table may be amended by notes on the approach chart. Such notes apply only to the particular approach category(ies) as stated. See legend page for description of components indicated below.

(1) ILS, MLS, and PAR

Inoperative Component or Aid	Approach Category	Increase DH	Increase Visibility
MM*	ABC	50 feet	None
MM*	D	50 feet	¼ mile
ALSF 1 & 2, MALSR, & SSALR	ABCD	None	¼ mile

*Not applicable to PAR

(2) ILS with visibility minimum of 1,800 or 2,000 RVR.

MM	ABC	50 feet	To 2400 RVR
MM	D	50 feet	To 4000 RVR
ALSF 1 & 2, MALSR, & SSALR	ABCD	None	To 4000 RVR
TDZL, RCLS	ABCD	None	To 2400 RVR
RVR	ABCD	None	To ½ mile

(3) VOR, VOR/DME, VORTAC, VOR (TAC), VOR/DME (TAC), LOC, LOC/DME, LDA, LDA/DME, SDF, SDF/DME, RNAV, and ASR

Inoperative Visual Aid	Approach Category	Increase MDA	Increase Visibility
ALSF 1 & 2, MALSR, & SSALR	ABCD	None	½ mile
SSALS, MALS & ODALS	ABC	None	¼ mile

(4) NDB

ALSF 1 & 2, MALSR, & SSALR	C	None	½ mile
	ABD	None	¼ mile
MALS, SSALS, ODALS	ABC	None	¼ mile

Fig. 14-2. *The Inoperative Components Table found in the front of every NOS Instrument Approach Procedures booklet.*

Inoperative Components

4. *Check inoperative component changes in minimums.* See FIG. 14-2. If any components of an approach listed on this table is out of service, the minimums may have to be increased. The table is published in the front of every NOS approach chart booklet. Definitions for MM, ALSF, MALSR, etc. are also in the front of the NOS booklets.

For nonprecision approaches, the inoperative component adjustment is relatively simple. As seen in (3) and (4) above, no increase is necessary in the MDA for inoperative components. However, there are increases of ¼ and ½ mile in visibility minimums when certain approach lights and runway lights are inoperative.

Check the lighting legend on the inside back cover of the NOS booklet (FIG. 14-3) against the airport diagram to see if the airport has any lighting systems affected by the inoperative components table.

This looks a little intimidating at first, but an instructor can help you sort things out. If you make a habit of checking the destination against the inoperative components table and the lighting legends every time you file IFR, you will soon be able to handle this problem quickly and easily. You will also broaden your understanding of the roles played by the various components, and what the wide variety of approach and runway lights look like.

The inoperative components check for Oxford reveals nothing that changes any minimums, which remain at 1,180-1 straight-in, 1300-1 circling.

5. *Make adjustments required by the fine print.* Here is where Oxford throws a zinger at the unwary pilot.

As noted above in item 1, MDAs must be increased 160 feet at Oxford if a local altimeter setting is not available. And there is a good chance this local altimeter setting will not be available and the White Plains altimeter setting must be substituted.

Oxford altimeter might be available on CTAF, which stands for *C*ommon *T*raffic *A*dvisory *F*requency. This could be the frequency for a part-time control tower after it is closed for the evening, it could be a Flight Service Station (FSS) located at the airport, or it could be a Unicom. Because Oxford is an uncontrolled airport and has no FSS facility, the CTAF is the Unicom frequency, which is manned by the FBO. If you show up when everyone at the FBO has gone home, you won't get an altimeter setting.

Be on the safe side and go ahead and add 160 feet to the Oxford MDAs in the flight planning. If it turns out that you can get an Oxford altimeter setting from the FBO, it will be a simple matter to glance at the approach chart and pick out the published MDAs. It would not be a simple matter to add 160 to the two minimums at the last minute when trying to descend to minimums, time the final approach segment, talk to ATC, listen on CTAF, and do a pre-landing check!

Better to add the 160 feet in the quiet of the planning room, producing adjusted minimums of 1,340-1 and 1,460-1. If you use Jeppesen charts, this calculation has already been made (FIG. 14-4.)

Fig. 14-3. *Approach Lighting Systems codes and descriptions found on the inside back cover of every NOS Instrument Approach Procedures booklet.*

Fig. 14-4. *Jeppesen version of the NDB RWY 36 approach at Waterbury Oxford, Connecticut.* Reproduced with permission of Jeppesen Sanderson, Inc.

6. *Add your altimeter error.* As discussed in Chapter 6, always *add* the altimeter error, regardless of whether it is plus or minus.

For purposes of illustration, let's say that you find an error of 30 feet when you check the altimeter. This will give adjusted MDAs at Oxford of 1,370-1 for a straight-in approach and 1,490-1 for a circling approach.

To summarize the step-by-step method of analyzing minimums:

1. *Read the fine print.*
2. *Check the height of obstacles.*
3. *Pick the correct minimums for airplane category and type of approach.*
4. *Check adjustments for inoperative components table.*
5. *Make adjustments required by fine print.*
6. *Add your altimeter error.*

OPERATION BELOW MDA

The reason I urge students to spend time analyzing the MDAs for the approaches they expect to make is because a pilot *cannot descend below an MDA at any time during a nonprecision approach unless certain very specific requirements are met.* These are stated in FAR 91. 116 (c). The regulation can be summarized:

No pilot may operate an aircraft below the authorized MDA (or continue an approach below the DH) unless:

(1) The aircraft is continuously in a position from which a descent to a landing on the intended runway can be made at a normal rate of descent using normal maneuvers.
(2) The visibility is not less than that prescribed for the approach being used.
(3) At least one of the following visual references for the intended runway is distinctly visible to the pilot:
    (i) The approach light system, including the red terminating bars or the red side row bars.
    (ii) The landing threshold.
    (iii) The threshold markings.
    (iv) The threshold lights.
    (v) The runway end identifier lights (REIL).
    (vi) The visual approach slope indicator (VASI).
    (vii) The touchdown zone or touchdown zone markings.
    (viii) The touchdown zone lights (TDZL).
    (ix) The runway or runway markings.
    (x) The runway lights.
(4) The visual descent point (VDP) has been reached on approaches that incorporate this feature.

Visual Descent Points (VDPs) are very helpful, but they are not common, and they require DME. A VDP is a DME distance on the final approach course of a nonprecision approach from which a normal descent can be made to a landing, provided one of the visual references above are visible.

VISIBILITY MINIMUMS REQUIRED FOR LANDING

Even if you can see the runway (or one of the other visual references listed above) as you approach the field at MDA, you may not legally make a landing if the visibility is less than that prescribed for the instrument procedure being used. This is the regulation, and it is stated in FAR 91.116 (d).

It is visibility, not ceiling, that determines whether or not you can land. MDA establishes the altitude below which you cannot descend unless you have one of the prescribed references in sight. Visibility tells you whether or not you can legally land once you have one of those prescribed references in sight.

Visibility is expressed in miles and fractions of miles, or in feet of RVR (Runway Visibility Range).

Visibility is the prevailing horizontal visibility near the surface as reported by an accredited observer. ATC tower controllers are qualified to report visibility.

RVR is measured by a transmissometer located alongside a runway. If a runway has a transmissometer, the visibility minimums listed on the approach chart will be expressed as a two-digit figure representing feet of RVR. If there is no transmissometer, the visibility will simply be expressed in miles, as is the case with the Oxford NDB approach.

If the transmissometer is out of service, the published RVR minimums must be converted to miles and fractions of miles according to the table on the inside front cover of the NOS booklet (FIG. 14-5).

RVR/Meteorological Visibility Comparable Values

The following table shall be used for converting RVR to meteorlogical visibility when RVR is not reported for the runway of intended operation. Adjustment of landing minima may be required – see Inoperative Components Table.

RVR (feet)	Visibility (statute miles)	RVR (feet)	Visibility (statute miles)
1600	$1/4$	4500	$7/8$
2400	$1/2$	5000	1
3200	$5/8$	6000	$1\frac{1}{4}$
4000	$3/4$		

Fig. 14-5. *Table for converting RVR to miles and fractions found on the inside front cover of every NOS Instrument Approach Procedures booklet.*

Missing RVR also increases the minimum visibility for some ILS approaches. This can be seen on the Inoperative Components table (FIG. 14-2.)

If making an approach at an uncontrolled field, you the pilot must decide if the visibility meets the requirements. Check the length of the landing runway; it makes a good reference for estimating visibility. Is it a mile long (5,280) feet)? Or half-a-mile (2,640) feet? Or a mile-and-a-half ? If you can see to the end of a mile-long runway as you begin your descent, you may legally land when the visibility minimum is one mile. But if you can only see partway down that runway, a landing may be illegal, and you should execute a missed approach. (This is a subject of much controversy and misunderstanding in the aviation community.)

If you are approaching a controlled field, the tower will inform you of the visibility. When it drops below the prescribed visibility minimums, the runway involved—or the entire airport—may be closed to landing traffic.

MISSED APPROACH PLANNING

While you are still in the planning room, assume that you will be required to make a missed approach and plan for it. Review it again before making the approach.

The Missed Approach Point (MAP) is no time to fumble for the approach chart and try to figure out what you're supposed to do. All attention must be riveted on controlling the airplane during the first few moments of a missed approach. You must add full power, stop the descent and initiate a climb, raise the flaps and gear, and maintain a steady course. Diverting attention to the fine print of the missed approach procedure at this time could start a chain of events leading to a collision with an obstacle, or the ground.

So plan ahead. I find that students have little trouble coping with a missed approach if they break it into five phases:

1. The first phase is the *transition* to a stabilized climb. You add full power, stop the descent, raise the flaps in increments, get the gear up, and initiate a normal climb straight ahead. (You have been practicing minimum controllable airspeed under the hood; now all that practice becomes very valuable.)
2. Second is the *climb* phase. Do you climb straight ahead or make a climbing left turn or a climbing right turn? The missed approach for Oxford prescribes a climbing left turn (FIG. 14-1). What is the level-off altitude? At Oxford it is 2,300 feet.
3. The third phase is *en route* to the holding fix. What is the holding fix? Is it a facility you already have tuned in, as at Oxford? Or is it a VOR fix that might require resetting frequencies and OBS numbers? Do you proceed direct? Do you have to intercept a bearing or radial to get to the fix?
4. Fourth is the *holding* phase. What type of pattern entry will you use? What outbound heading do you turn to when you reach the fix? Write it in big numbers on the approach chart.
5. Fifth is holding *departure*. Plan for two alternatives: returning for another instrument approach or diverting to the filed alternate. When approach control asks "What are your intentions?" have your mind made up and respond promptly what you intend to do, including an abbreviated flight plan containing route and altitude to the alternate, if that's what you decide to do.

You cannot depart the missed approach holding pattern until cleared to do so either for another approach or to proceed to the alternate airport.

Once again, visualization is the key to success in working out the moves made on a missed approach. Visualization is also the key to success in making the basic approach itself. A good instrument approach, which always includes the missed approach procedure, begins the night before, along with your planning for the departure and en route phases. Mentally fly the approach step-by-step, or even better, walk through it by placing objects on the floor to simulate the airport and the approach and missed approach fixes.

As the final step in approach planning, run through a "MARTHA" check:

MA (Missed Approach procedures.)
R (Radios—nav and com frequencies and OBS settings.)
T (Times from FAF to MAP.)
H (Heading of final approach course.)
A (Altitude of MDAs, adjusted as discussed above.)

This abbreviated approach checklist will also come in handy in the air near the destination while preparing for the approach.

With a little practice, you will find that planning an approach takes far less time than reading about it!

NDB APPROACHES

I always introduce students to NDB approaches before VOR or ILS approaches. This may come as a surprise, but it really shouldn't. The two lessons that precede approaches in the syllabus are devoted to ADF procedures, so ADF is still fresh in the student's mind.

And, because VOR is the backbone of the federal airway system, most students start instrument training with far more experience with VOR than ADF. So I pay extra attention to ADF as we move through the course. By introducing NDB approaches before the others, I can make sure the student is skillful, confident, and comfortable with them. If NDB approaches are introduced later in the course, they may not get the attention they require.

Let's return to the NDB approach at Oxford (FIG. 14-1) and talk through the procedure, step-by-step. And, because this is the first approach I discuss in detail in this book, I will also introduce material on approach control, communications, and flight procedures that apply not only to NDB, but to approaches in general.

RADAR VECTORS

In the real world of IFR, you will be handed off from the ATC center controller to the appropriate approach controller at a comfortable distance from your destination.

You will frequently be cleared to a lower altitude just before or just after the handoff to approach control. Leaving an assigned altitude is one of the occasions for a *required* report whenever this occurs during an IFR flight. The readback to ATC will be like this:

Cessna five six Xray contact approach control one two zero point seven five, descend to three, leaving five.''

Remember to use our full call sign on initial contact with approach control.

Approach control will give you a time to "expect further clearance" or expect approach clearance for use in case of lost communications. The time they give will also help you plan the approach. If you don't get a "further" time, request it.

Approach control will issue vectors to intercept the final approach course (004° at Oxford) one to five miles outside the Final Approach Fix (CLERA NDB), where you will be "cleared for the approach." This will give you time to establish yourself on the final approach course before reaching the FAF, to slow to approach speed, and to prepare for the final descent and landing.

Treat radar vectors as commands. They are issued as required to provide safe separation for incoming traffic. Therefore do not deviate from the headings and altitudes issued by approach control.

Sometimes it becomes necessary for ATC to vector you *across* the final approach course for spacing or other reasons. This is not unusual at busy airports with a mix of slow traffic and high speed traffic. It is much easier to move you out of the way of a rapidly closing jet than to have it break off the approach. After the jet has passed, you will be vectored back to the final approach course with minimum disruption.

You will normally be informed when it becomes necessary to vector you across the final approach course. If you see that interception of the final approach course is imminent and you have not been informed that you will be vectored across it, question the controller.

Simply give your call number and "Final approach course interception imminent, request further clearance." You will be cleared either to complete the approach or to continue on present heading for separation from incoming traffic. Do not turn inbound on the final approach course unless you have received an approach clearance.

THE FULL APPROACH PROCEDURE

In the beginning of your intensive work on NDB approaches, however, skip the radar vectors and work the "full approach procedure" to become skilled in all elements of the approach.

Make this request for the full approach on initial contact with approach control after the handoff from center. You will probably be cleared direct to the FAF, in this case, CLERA NDB, and receive an expect further clearance or expect approach clearance time.

If "direct" to the FAF is within 10° of the final approach course (004°), go ahead and intercept the final approach course and proceed direct to the FAF. Approach control will expect you to do this and will clear you for the approach before reaching the FAF.

HOLDING PATTERN COURSE REVERSAL

If you can't line up with the final approach course and proceed direct to the FAF, plan to enter the holding pattern shown on the approach chart and return to intercept the final approach course. This is a "course reversal" in a holding pattern and it is a widely used method of lining up. When a holding pattern is used for this purpose, the racetrack pattern shown on the chart will be printed with a much darker and heavier line than other holding patterns (FIG. 14-4).

It is very important to the success of an NDB approach to get lined up on the final approach course as soon as possible with an accurate wind correction. If not lined up properly at the FAF, the chances of making a successful approach are jeopardized.

The holding pattern is one of two methods of "course reversal." The other method for lining up is the "procedure turn." I will discuss procedure turns in the VOR approach section of this chapter. You will see both methods of course reversal—the holding pattern and the procedure turn—widely used in both NDB and VOR approaches.

If the NDB approach you plan to practice has a procedure turn, by all means skip ahead to the VOR section and study procedure turns. Use the course reversal shown on the approach chart—either holding pattern or procedure turn—for obstacle clearance reasons. You do not have the option of substituting one for the other.

APPROACH SPEEDS

During the course reversal and in steady flight in a "full approach" or as you head toward the FAF with radar vectors, slow to approach speed and lower "approach flaps," usually one increment. One hundred knots is a comfortable approach speed for most light airplanes. This will usually be a 90-knots ground speed in typical winds.

The exact speed doesn't make too much difference as long as it is a comfortable speed that you can hold constant throughout the approach, including the descent to MDA after leaving the FAF.

If you are aware of faster traffic behind you, maintain cruise airspeed and keep the flaps up. Plan to make a high speed final approach at 110 knots, or even 120 knots if safe. Don't worry about coming in too fast and using up too much runway! If it is long enough for the jets behind you, that runway will be plenty long enough!

Complete the approach checklist as you head toward the FAF, and run through the MARTHA check again. All approach charts for your destination should be on the clipboard with the probable approach chart on top.

It is very helpful to clip approach charts to the yoke for quick reference throughout the approach and missed approach. Some airplanes come equipped with a yoke chart clip; you may also purchase one at aviation supply companies and many FBOs.

APPROACH COMMUNICATIONS

If you have to hold at the FAF, you must make another *required* report as you enter the hold, as follows:

"Approach Control, Cessna five six Xray, CLERA, entering hold, level at three."

As the "expect further clearance" time approaches, you can anticipate that approach control will either clear you for the approach, issue a revision of the time, or—at a controlled airport—hand you off to tower. If the latter is the case, you will be given the tower frequency.

At an uncontrolled airport, you will be asked how you plan to terminate the approach. You have three options: (1) land, (2) make a low pass and cancel IFR, and (3) execute a missed approach.

With the first option—a landing—approach control will tell you "report landing or landing assured." You will remain on the approach control frequency until advised "Frequency change approved." You must then switch to the CTAF frequency and report your position on the CTAF to alert other traffic to your position and that you are inbound on a specific instrument approach.

CTAF is also the frequency to get the altimeter setting and the runway in use.

When you report landing or landing assured to approach control, the IFR flight plan will be closed by ATC.

With the second option—low approach and cancel IFR—the IFR flight plan will be cancelled when you announce "cancel IFR" to approach control.

With the third option—missed approach—you return to the approach control frequency at the MAP. Executing a missed approach is the occasion for another *required* report. But you don't have to make this report as soon as you add power for the missed approach. Wait until you have stabilized in the climb and everything is under control, then report. Always remember aviate, navigate, communicate!

You can expect approach control to issue an abbreviated clearance for returning to the missed approach holding fix. It is not a good idea to try to copy a clearance while you still have your hands full controlling and cleaning up the airplane in the transition phase of a missed approach. Wait until you are in a stable climb, then contact approach control.

FLYING THE NDB APPROACH

So far we have discussed elements of the instrument approach that also apply to all nonprecision and precision approaches to one degree or another—radar vectors, approach speeds, course reversals, and terminating the approach. Now let's back up a bit and discuss how you will actually fly the full procedure in our chosen example, the NDB 36 at Oxford.

Proceed to the FAF, CLERA NDB, following the clearance from approach control. On your way to CLERA, slow to approach speed and review the MARTHA check.

On reaching CLERA, run through the "Five Ts" as you always do at a fix or making a change in course or altitude:

- Write down the *time* of arrival (ETA) at CLERA on the approach chart. You will also need to start timing the outbound leg.
- *Turn* to the appropriate heading for the holding pattern entry you have selected—direct, teardrop, or parallel.
- *Twist* is not necessary for an NDB approach. Instead, use this item as a reminder to adjust the volume on the identifier to hear it faintly in the background. Monitor the NDB identifier continuously throughout the approach to detect a failure of either the transmitter or receiver.
- Reduce *throttle* to 100 knots if you have not done so already.
- Arrival at a holding point *requires* a report. *Talk.* Say "Cessna five six Xray, CLERA at [time], three thousand (assigned altitude)." That's all you need.

WHEN TO DESCEND

When approach control has cleared you for the approach, you may begin a descent to the altitude prescribed on the Approach Chart—2,300 feet at Oxford—as soon as you depart CLERA outbound for the holding pattern entry. If you have not been cleared for the approach, you must remain at your assigned altitude—3,000 feet in this example—until approach control clears you to a lower altitude, or says "cleared for the approach." This reason for remaining at the assigned altitude is obvious because there may be other airplanes in a holding pattern below.

Make a normal, stabilized, constant airspeed descent. Slow the airplane to the approach speed you have selected, say 100 knots (if you haven't already done so). When stabilized, reduce power 100 rpm (or 1″ of manifold pressure) for each 100 feet per minute you want to descend. A reduction of 500 rpm—from 2,300 rpm to 1,800 rpm, for example — will produce a rate of descent of 500 feet per minute. A reduction of 5″ in manifold pressure will also produce a 500-foot-per-minute descent.

It's always a good idea to start a descent as soon as you are cleared to do so. The sooner you get down to the desired altitude, the more time you have to stabilize altitude, airspeed, and heading. This becomes very important as you descend to the MDA on the final approach course. If you have a large amount of altitude to lose, descend at 1,000 feet per minute until 1,000 feet above the desired altitude, then reduce the rate of descent to 500 feet per minute.

Make the holding pattern entry. As you head inbound toward CLERA, intercept the inbound course (004° and begin bracketing to establish a reference heading that will correct for the wind and maintain an inbound course.

On reaching the FAF there are several important steps to take:

- Start timing the final approach segment.
- Adjust heading as necessary.
- Make sure the volume is correctly set so you can faintly hear the identifier throughout the approach.
- Reduce power 500 rpm (or 5″ of manifold pressure) to begin a 500-foot-per-minute descent to the MDA.
- At a controlled airport, contact tower if you haven't already done so. At an uncontrolled airport, report on CTAF passing the FAF. Always report your position and intentions on CTAF to alert local traffic of your position. Make frequent additional reports on final as needed.

TIMING THE APPROACH

I recommend using a stopwatch to time the final approach segment from FAF to MAP, or a digital timer on the instrument panel that you can start as you pass the FAF. A stopwatch (or digital timer) is also handy for timing the legs of a holding pattern.

You can also use the sweep-second hand of the conventional clock to time the approach. But many students find it confusing trying to keep track of how many minutes have passed on a long final approach segment. On some long finals, the time from FAF to MAP can be more than five minutes.

As you concentrate on maintaining the MDA and the final course, it is easy to forget how many times the sweep-second hand has gone around. (Is it three? No, that was last time. Must be four. But it's taking so long! Maybe I've already gone five minutes!)

Eliminate the confusion altogether. Buy a timer at the beginning of your IFR training and use it on every approach. You will soon find that accurate timing ceases to be a problem.

The times from FAF to MAP at different speeds are located in the lower right corner of the NOS Approach Charts. These times are based on no-wind conditions, so you must adjust them for your estimated ground speed. If you are good with numbers you can interpolate and quickly find the time to match your ground speed.

If you make this calculation a part of the MARTHA check as you approach the airport, it will save a lot of fumbling when you reach the FAF.

None of these calculations will be accurate if you cannot fly a constant airspeed as you descend, then level off at the MDA.

Tip: Use the next higher speed when timing the FAF to MAP segment when you know the ceiling and visibility to be well above minimums, say 600 overcast and 2 miles.

FINAL APPROACH COURSE

Another problem I see frequently when giving flight tests for the instrument rating is pilots getting so disoriented on the final approach course after passing the FAF that they cannot find the field.

The basic problem here is poor training or lack of practice tracking and bracketing the NDB. If you understand and practice the procedures discussed in Chapter 13, "The NDB Unmasked," you should have little difficulty in this phase of the approach.

As you head toward the FAF you should have enough time to bracket the inbound course and determine a reference heading that will correct for wind and maintain that course. At station passage, don't chase the needle. Maintain the reference heading outbound from the FAF until the needle settles down and you have completed your "Five Ts" checklist. After that you can make minor adjustments in the reference heading if necessary.

If you have an accurate reference heading when passing the FAF, all you have to do is maintain that reference heading and you will see the landing runway when the time is up—ceiling and visibility permitting.

There are times, however, when even the pilots are unable to establish an accurate reference heading as they fly toward the FAF. The wind might be changing rapidly or approach control might turn you inbound so close to the FAF that there is not enough time to get established on course.

Do the best you can heading inbound in this situation. Then at station passage turn immediately to the inbound course, wait 10 seconds for the needle to settle down, note the number of degrees the needle is off to the left or right, and then re-intercept and bracket the outbound course using the procedures described in Chapter 13. That will get you back on course before you get to the MAP.

MISSED APPROACHES

There may be times when even this won't work. If you cannot establish yourself on the final approach course for any reason—or if you have lost track of the timing—you must execute an early missed approach.

The procedure for an early missed approach is different from a missed approach at the MAP. In an early missed approach, add full power, clean up the plane, establish a normal climb, report to ATC, and transition to the published missed approach procedure. Above all, *do not make a turn until you have reached the MAP.* You are not guaranteed the full obstacle clearance associated with that approach if you depart from the final approach course.

A decision to make a normal missed approach at MAP is based on several variables.

First, what would cause a missed approach? Very often missed approaches are required because you didn't get down to MDA and were unable to establish the visual references spelled out by FAR 91.116 (c) for a further descent. A missed approach is also required if the visibility is below that required for the approach. Or perhaps you weren't lined up properly and just caught a glimpse of a corner of the airport as you flew by.

You might also have to make a late missed approach if you have the required visual references and the visibility needed to land but find, after you begin the descent for landing, that you cannot land for some reason, like another airplane on the runway.

Remember the five phases of a missed approach:

1. *Transition* to a stabilized climb.
2. *Climb* straight ahead or in a climbing left or climbing right turn, as prescribed in the published missed approach procedure.
3. *En route* to the holding fix.
4. *Holding* at the designated holding fix.
5. *Departure* from the holding fix for another approach or to an alternate.

There is no limit, other than the amount of fuel on board, to the number of approaches you may attempt. If you missed the approach for reasons other than weather—poor line up for example—go back and try again. But if you reached MDA and the weather was below minimums, or if the weather was obviously deteriorating, the smart move would be to proceed to the alternate.

CIRCLING APPROACHES

You are on the final approach at MDA and as you run out of time hopefully you run out of clouds and the airport is in sight. You do a landing check passing through 500 feet above the airport, lower full flaps, and land.

But somewhere prior to completing the landing check and lowering full flaps, you might find that you have to land on a different runway. If you are landing at a controlled airport, the tower will make that decision and clear you to "circle and land Runway [as assigned]."

At an uncontrolled field, however, you must make the decision yourself. There are several variables that affect this decision.

In some cases there is no choice. Many approaches are not sufficiently aligned with the runway to permit a true straight-in approach. When a procedure does not meet the criteria for straight-in approaches, it is designated A, B, C, and so on (NDB-A, NDB-B, NDB-C, etc.), and no straight-in minimums are published. An example of this is the NDB-A approach at Perkasie/Pennridge, Pennsylvania (FIG. 14-6). The final approach course is 181° and the only runway at Perkasie is 8-26. So a circling approach is the only alternative available.

Conditions at the time of reaching the MAP may dictate a circling approach. If the crosswind is too great for the straight-in runway, for example, you should choose a landing runway that is closer into the wind, if one is available. If there isn't a better runway, execute a missed approach.

Sometimes you learn that airplanes in the landing pattern ahead of you are using a different runway. Learn the active runway from unicom, or from other airplanes in the pattern as they report their positions on the CTAF. If a different runway is in use when

Fig. 14-6. *Typical circling approach procedure when runway is not aligned with final approach course.*

you arrive, you will have to make a circling approach and fit into the traffic pattern. If there is traffic in the landing pattern, it better be VFR, so you should break out in VFR conditions well above the circling MDA. There's nothing that says you can't fly a circling approach higher than the circling MDA as long as you are clear of clouds.

You may get right down to the straight-in MDA before deciding to make a circling approach. Maybe another airplane taxis out and dawdles on the runway just as you are about to land. Add power and go around, just as you would under VFR conditions.

But you are still IFR so you must circle around again for another attempt at landing. In this case you must climb back up to the published circling minimums, or traffic pattern altitude, in order to continue.

If that puts you back in the clouds, you will have to execute a missed approach. You must keep the runway of intended landing in sight at all times during a circling approach, or execute a missed approach.

Circling Approach Patterns

The FAA recommends the circling approach patterns shown in FIG. 14-7. Pattern A can be used when the final approach course intersects the runway centerline at less than a 90° angle and you see the runway clearly enough to establish a base leg.

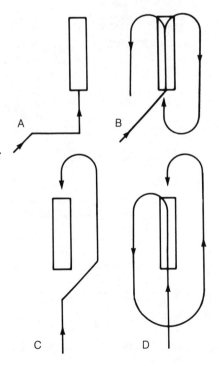

Fig. 14-7. *Circling approach patterns. Use A when final approach course intersects runway centerline at less then 90°; use B if you see runway too late to fly pattern A: use C to land in opposite direction from final approach course; use D if you see the runway too late to use pattern C.*

If you see the runway too late to fly Pattern A, circle as shown in B and make either a left downwind or a right downwind. Fly pattern C if it is desirable to land opposite the direction of the final approach course and the runway is seen in time for a turn to the downwind leg. If the runway is sighted too late for a turn to the downwind as shown in C, fly pattern D.

So far this all sounds very reasonable. However, the circling minimums may be *as much as 500 feet lower than a VFR traffic pattern* for the same runway flown at 1,000 feet above field elevation. Some people call the circling maneuver legal scud running. It takes some very careful maneuvering to make a safe approach and landing from a low altitude.

Furthermore, the circling minimums guarantee an obstacle clearance of *only 300 feet* within the "circling approach area." This is a very small area, and you must remain within it.

See FIG.14-8. TERPS describes how the circling approach area is constructed. The circling approach area is based on the same aircraft approach categories A, B, C, D, and E that appear in the minimums sections of instrument approach charts. For category A, which most of us use for instrument training, the circling area has a radius of only 1.3 nautical miles from the end of each runway.

FIGURE 14-8 shows how arcs drawn from these radii outline the area in which obstacle clearance is provided. *Outside this area, there is no obstacle clearance protection.* Keep the circling approach pattern within the safe area. Use the runway length to help visualize 1.3 nm.

You are required to perform a circling approach on the instrument flight test; master the skills necessary to carry out this maneuver at circling MDA. However, circling approaches are imprecise and may be dangerous.

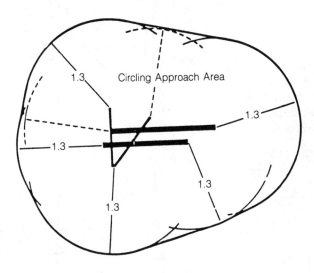

Fig. 14-8. *Circling approach area in which obstacle clearance is provided.*

Instructor Note. Practice, as always, is the best way to build confidence in circling approaches. In VFR conditions, make it a routine to terminate one instrument approach on every flight with a circling approach at the VFR pattern altitude so the student will learn to sequence with other traffic.

In actual IFR, take advantage of every opportunity to have the student make circling approaches at circling MDA at an uncontrolled airport. The tower at a busy airport probably won't let you do this just for practice, but will insist that you land. You can make an instrument approach and continue with a circling approach until the student is in a position to land. Execute a missed approach at that point and obtain ATC approval to return for more practice without cancelling the IFR flight plan.

Be sure to brief the student about your intentions. He won't get much out of the practice if he's totally confused about what's going on!

NDB ON AIRPORT

Many NDB approaches are based on the compass locator of an ILS approach, but some NDBs are located right on the airport, as seen in FIG. 14-9, the NDB RWY 22 approach at Easton, Maryland, a busy field on Maryland's popular Eastern Shore.

An NDB approach with the NDB on the airport is a very simple approach. Proceed to the NDB, which is both the initial approach fix (IAF) and the MAP. (There is often no FAF when the NDB is located on the field.) Turn outbound on the indicated course, in this case 028° the reciprocal of the inbound course.

A procedure turn is indicated rather than a course reversal in a holding pattern. I will discuss procedure turns in detail in the section on VOR approaches, so skip ahead to that section if you intend to work with an NDB-on-airport approach during this phase of training.

Note that the fine print in the profile section of the approach chart says ''Remain within 10 NM.'' This means that you must complete the procedure turn within 10 nautical miles of the NDB.

How far is 10 nm? Work it out on the circular slide rule while in the planning room. At a ground speed of 90 knots it takes six minutes 40 seconds to cover 10 nm; at 100 knots ground speed it takes six minutes exactly. So you should complete the procedure turn before the times expire, depending on your ground speed. If you stray beyond 10 nm, obstacle clearance is not guaranteed.

The normal procedure is to fly outbound for one minute, depending on the wind, then begin the procedure turn. Descend to procedure turn altitude (1,600 feet in this case) while heading outbound and during the procedure turn.

Do not descend below 1,600 feet until you intercept the inbound course, 228°. Again, this is for obstacle clearance reasons. After intercepting the inbound course, you are free to descend to 660 feet, which is the MDA for both straight-in and circling approaches at Easton.

Fig. 14-9. *NDB approach with station located on the airport.*

Note that no times are given. That's because there is no FAF. The MAP is the NDB itself. So there is no need to time the inbound leg. Just stay on the inbound course of 228° at the MDA of 660 feet until station passage occurs. If you don't have the visual references you need to descend below MDA when the needle reaches the 90° position, you must execute the missed approach.

Planning steps, communications, MARTHA and Five T checks, approach speeds, and missed approach phases will be the same for an on-airport NDB approach and an NDB approach with a FAF some distance from the field.

VOR APPROACHES

One way to gain a good overview of approaches is to examine them as short cross-country flights. A VOR approach is nothing more than a miniature cross-country. You proceed from the last en route or feeder fix to the IAF, to the FAF, and then to the MAP following predetermined courses, making turns and changing altitudes as required.

It may take two to five minutes to get from the en route or feeder fix to the IAF and then another five to 10 minutes to arrive at the FAF. Remember that you will be tracking inbound and outbound by VOR during the approach just like you track VORs inbound and outbound on a cross-country.

The planning, communications, MARTHA and Five T checks, approach speeds, and missed approach phases remain the same for VOR approaches as for NDB approaches.

Let's work our way step-by-step through the VOR-A approach (FIG. 14-10) at Poughkeepsie/Dutchess County, New York. Once again, request the full procedure in order to get the maximum training out of this exercise.

There are two items of special interest in this approach. One, it contains a "dog leg," and two, it has a procedure turn rather than a course reversal in a holding pattern.

Proceed to the IAF, Kingston VOR, do the Five Ts and turn outbound and get established on the 037° radial, which is the reciprocal of the inbound course.

PROCEDURE TURNS

Fly outbound on the 037° radial for one minute and make a 45° turn to the left as shown on the chart. Reset the OBS to the inbound course, 217°. This is the beginning of the procedure turn, an easy, reliable method of course reversal that will return to the inbound course with fewer corrections.

Fly outbound on the 352° heading for one minute, adjusting for the wind, then make a 180° turn. This will establish you on the 172° course. Intercept the inbound course using the method described in Chapter 10. Hold the 172° intercept heading until the needle is about three quarters of the way from full-scale deflection toward the center, then begin a turn to the inbound course. Correct for the wind and establish a reference heading that will hold the inbound course to the station.

Amdt 10 88098

VOR-A

AL-286 (FAA)

POUGHKEEPSIE/DUTCHESS COUNTY (POU)
POUGHKEEPSIE, NEW YORK

BOSTON CENTER
120.2 256.9
DUTCHESS COUNTY TOWER *
124.0 (CTAF) 357.4
GND CON
121.8
ATIS 126.75
POUGHKEEPSIE RADIO
122.4

10 NM

R-037

172°
352°
217°

PAWLING
114.3 PWL
Chan 90

3000°
250°
(11.8)

1570

1400

037°
360   621 752
487±
600   602±
242°
513
428
286

IAF
KINGSTON
117.6 IGN
Chan 123

1365

1456

1336

R-242

2000

1036

1740

.1326

MSA IGN 25 NM
190°
4600   3500
100°   280°
3300   2500
010°

530' terrain 1.2 NM NW of airport.
567' tower 1.5 NM NE of airport.

△1500

MISSED APPROACH
Climb to 1500 via IGN R-242
within 5 NM then climbing right
turn to 2800 direct IGN VORTAC
and hold.

VORTAC

Remain
within 10 NM

037°
2800
217°
242° 1600

3.3 NM

ELEV 166

Rwy 33 ldg 2872'

242° 3.3 NM
from FAF

198△

197△

265

215

176

CATEGORY	A	B	C	D
CIRCLING	760-1 595 (600-1)	820-1 655 (700-1)	820-1¾ 655 (700-1¾)	820-2 655 (700-2)

MIRL Rwy 15-33
HIRL Rwy 6-24
REIL Rwys 24 and 33

	FAF to MAP 3.3 NM				
Knots	60	90	120	150	180
Min:Sec	3:18	2:12	1:39	1:19	1:06

VOR-A

41°38'N-73°53'W

POUGHKEEPSIE, NEW YORK
POUGHKEEPSIE/DUTCHESS COUNTY (POU)

Fig. 14-10. *VOR approach with a "dog leg."*

It should be noted that there is no "right" way to make a procedure turn. Nowhere is it written that you must use the 45° procedure published on the approach charts. All that is required is that somehow you get turned around and headed back on the inbound course within the mileage limit published on the chart, usually five or 10 nm.

But it makes good sense to use the 45° procedure published on the approach charts. The 45° headings are printed on the chart, so there is no guesswork about headings. And the 45° method will enable you to intercept and get established on the inbound course quickly and easily.

The distance limit on the procedure turn is not there just to keep you from wandering too far. It is a limit beyond which a minimum obstacle clearance altitude of 500 feet cannot be guaranteed at the procedure turn altitude.

Let's see this in practice. Suppose you had a ground speed of 120 knots heading outbound and had to remain within five nautical miles for obstacle clearance. How many minutes would it take to reach the five nautical mile limit? Just 2 minutes and 30 seconds.

If you include this calculation in preflight planning, you won't have to fumble for it at the last minute, outbound for the procedure turn. As a rule of thumb, a one-minute outbound leg will keep you within a five mile limit, unless there are unusually strong winds.

As noted before in the discussion of NDB approaches, a holding pattern may be prescribed in lieu of a procedure turn. If an approach has the note "NoPT," no procedure turn is required and *you may not execute it* without clearance from approach control. A few VOR approaches state flat out, "Procedure Turn NA"—not authorized. Don't even think about it! There is probably a big mountain right where you would normally expect to make a procedure turn.

You may commence a descent to the procedure turn minimum altitude as soon as you pass the IAF. The procedure turn minimum altitude is 2,800 feet on our example, the Poughkeepsie VOR-A approach (FIG. 14-10). Do not go below the procedure turn minimum altitude until established on the inbound course. "Established" means a "live" needle, not necessarily centered. Once established you may descend to the FAF minimum altitude, 1,600 feet in this case.

On reaching the FAF, begin a descent to the MDA for the approach. Get down to the MDA as quickly as you can comfortably to give yourself the maximum opportunity to see the airport and pick out the landing runway. Make a constant airspeed descent so you do not throw your timing off.

Normally after completing a procedure turn you can expect to fly a straight line course to the FAF and then on to the MAP. But not the approach at Poughkeepsie. On reaching the FAF, make a right turn to 242° and proceed toward the MAP on this new course. That's why this is designated an "A" approach; that dog leg does not meet the criteria for a straight-in approach, even though you may end up lined up for a straight-in for Runway 24, if you fly a perfect approach!

You will be very busy at the FAF as you run through the Five Ts. You must start *timing* your final approach leg; *turn* to intercept 242°; *twist* the OBS to 242°; *throttle* back 500 rpm (or five inches of manifold pressure) to begin a 500 feet per minute descent at your approach speed; and *then talk* to approach control. The report will be "Cessna five six Xray, Kingston inbound." Aviate, navigate, communicate!

The key to coping with complications like this "dog leg" is to spot them while planning the flight, then talk through the approach until you understand clearly the course, turns, descents, and reports. Again, think of it as a miniature cross-country and be sure to include the missed approach procedure as part of the cross-country.

Fortunately, you won't encounter too many "dog leg" VOR or NDB approaches. But you must be prepared to handle them.

A more common variation on the VOR approach has the VOR located on the field. This is the case at Bridgeport, Connecticut (FIG. 14-11). This airport is located on a point of land jutting out into Long Island Sound. The only place approach facilities *could* be located is right on the field, or they would be underwater. When a VOR is located on the field, the MAP is reached when the TO-FROM indicator flips to FROM.

The Bridgeport VOR RWY 24 approach also has another frequently seen feature. Many of the fixes are VOR intersections, including the IAF at Milum and the missed approach holding fix at Stane. The 121° radial from Carmel (which you already have on the No. 2 VOR for the other fixes) can also be used as a double check for the MAP. As you talk through an approach like this in the planning room, consider writing out a separate sequence of OBS settings for all these fixes on the flight log. Be sure to include the other intersections shown on the approach chart (such as Weems on the Bridgeport VOR 24 approach chart) in case approach control specifies them in a clearance.

DME AND DME ARC APPROACHES

Proficiency in the use of DME (distance measuring equipment) is not a requirement of the Instrument Rating Practical Test Standards. But ATC expects you to be competent and able to use any equipment in your airplane. If you file equipment Code A (DME and transponder with altitude encoding capability), ATC will issue clearances with DME points in them. Be prepared to make DME approaches, some of which may surprise you if you haven't practiced them.

DME indications appear sometimes on VOR approach charts as an aid to making a conventional VOR approach. However, if DME is not included in the name of the approach procedure—if the name of the approach is simply VOR RWY 28, for example—then DME distances are just aids, and DME is not required for the approach.

On the other hand, if DME is included in the name of the approach procedure, as in FIG. 14-12, you must have DME to use the approach, unless ATC agrees to call out the DME fixes.

Note that the two IAFs for the VOR/DME RWY 15 approach at Johnstown/Cambria County, Pennsylvania, are located where two Johnstown VOR radials intersect the 10 mile DME arc. Then you fly the 10 DME arc around to intercept the 326° radial and turn inbound. The FAF is DME 4 on the inbound 146° course.

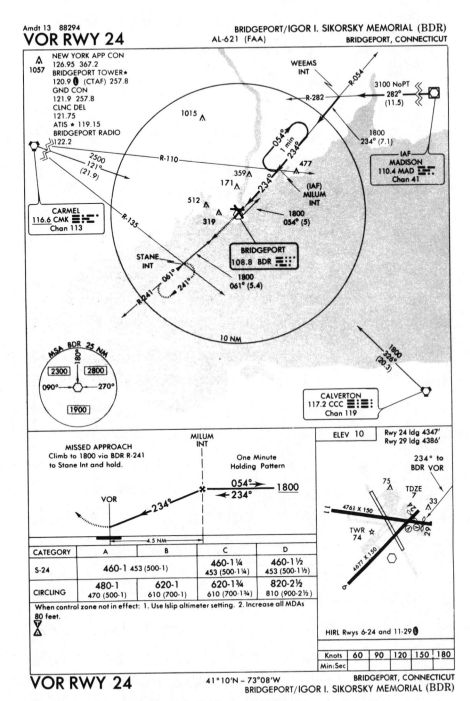

Fig. 14-11. *VOR approach with fixes at VOR intersections.*

Fig. 14-12. *Procedure for flying VOR/DME RWY 15 at Johnstown, Pennsylvania. Intercept radial 074 at (1), turn onto arc at (2), turn to inbound course at radial 326 (3).*

198

Flying a DME arc is not as difficult as it looks on the approach chart. You won't have to do the impossible and fly a smooth, continuous, perfect arc. Instead, fly a series of short, straight tangents to the arc, just as you did on a VOR Time/Distance check. These short tangents will keep you close to the 10 miles specified.

FLYING THE DME ARC

Let's work our way through the Johnstown VOR/DME RWY 15 approach, arriving from the northeast.

First, intercept the 074° radial and turn inbound (1) on the reciprocal bearing to the station, 254°. As the DME mileage clicks off, anticipate a turn at 10.5 DME. This will enable you to lead the turn onto the 10-mile arc by half a mile.

At 10.5 DME turn right 80° (2) to a heading of 334° (FIG. 14-12). When you complete the turn you will be on a tangent to the arc at a distance of 10 miles, or very close to it. Rotate the OBS 10° *opposite* to the direction of the first 90° turn—left in this case to a setting of 064°. As you continue on the 344° heading, the DME mileage will begin to increase. FIGURE 14-13 shows how the tangents are flown around the Johnstown arc.

Remember that when you set radials on the OBS, the CDI needle will start out on the *same side as the station* and move to the opposite side as you pass the radial. In this case, the CDI needle will move from left to right as you approach and pass the 064° radial.

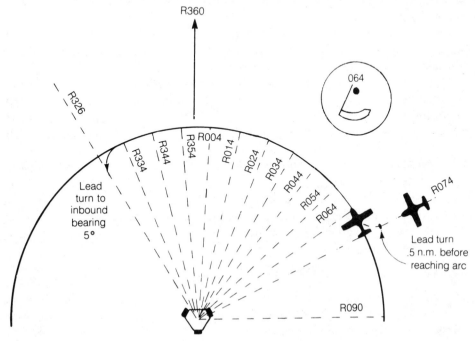

Fig. 14-13. *Details of flying the 10 DME arc at Johnstown. Compare with Fig. 14-12.*

When the needle centers, turn left 10° to a new heading of 324°. Reset the OBS 10° to read 054°. In a no-wind condition, the DME distance will decrease to 10 miles after the turn, then begin to increase again as you fly the tangent.

Continue with these 10° heading and OBS changes as you track around the arc toward the 326° radial. Does this remind you of the Time/Distance checks discussed previously?

Naturally, the wind will tend to blow you toward the station or away from it, depending on its direction. If you find the DME distance increasing, you are being blown away from the station. Make the next 10° heading change sooner, or make the turn more than 10°. This will bring you back inside the curve.

If the DME distance decreases, you are being blown toward the station. Reset the OBS for 20°. Make the next heading change 10° as usual. You will fly a longer tangent before the needle centers again. This will correct for the wind blowing toward the station.

Lead the turn onto the inbound course by five degrees. In the Johnstown example, you want to establish a 146° course inbound to the station. After passing R334, set the OBS to 151° to lead the inbound course (3). Do the Five Ts and start the turn inbound when the needle centers. After completing the turn, reset the OBS to 146° and track this course inbound.

Note how DME distances are used to fix the FAF at Hinks intersection and the MAP at .6 DME. Normally, when the VOR is located on the field, the MAP occurs at station passage. But in this case, you must execute a missed approach before reaching the VOR to avoid obstacles.

With the Johnstown VOR/DME RWY 15 approach, the DME arc is used to position you on a conventional VOR course, and for the FAF and MAP. In some cases, believe it or not, the DME arc *is the final approach course.*

(See FIG. 14-14, the VOR DME RWY 14 at Baltimore/Martin State, Maryland.) This is a fairly simple approach to fly, despite the way it appears on the chart. You intercept the Baltimore 331° radial, fly inbound to the 14.7 DME arc, turn left 90°, then begin making 10° tangents all the way around until the airport lies ahead. The various radials provide the descent points and the MAP.

It should be clear at this point that there is nothing conventional about a DME approach. Every one of them is different, sometimes radically so. But if you have DME aboard and take the time to practice DME approaches and get acquainted with them, you will gain access to a large number of approaches that you might otherwise not be able to use. (Believe it or not, DME arcs are easier to fly than to explain in written text.)

TIPS ON FLYING APPROACHES

The successful outcome of the approach is usually assured by thorough preplanning, by carefully studying the approach that will probably be used, and by having all approach charts for that airport readily available.

Know instantly where to look for all significant items on the approach chart.

Fig. 14-14. *VOR/DME at Baltimore/Martin State, Maryland, with DME arc as final approach course.*

Be prepared for the next step of the approach. Think ahead about the segment you are about to fly.

Don't try to comprehend or digest the entire approach chart all at once.

Always be prepared for a possible missed approach. "Gotta' landitis" prevents some pilots from growing older!

Keep the approach technique simple.

Slow to approach or holding speed before commencing the approach or during course reversal, and lower approach flaps.

Fly "by the numbers" at predetermined airspeeds and power settings to attain a trimmed configuration.

Determine the wind correction before reaching the FAF and fly the reference heading ±2° to 5° to maintain the desired track.

Perform a pre-landing check prior to reaching the FAF. Lower the landing gear at the FAF or make a power reduction in a fixed-gear airplane. Then note the time over the FAF, check heading and turn as necessary, change OBS if required, and report to the controlling facility.

Know where you are at all times!

15
Approaches II: ILS, Localizer, and Radar

THE ILS APPROACH IS THE MOST PRECISE APPROACH AVAILABLE TO THE GENERAL aviation instrument pilot. It is also the easiest to master. It must be easiest, because very few instrument students seem to have problems with it!

This might seem like a puzzle at first because the ILS approach is fairly complex and it requires a high degree of skill for heading and altitude control. What happens, I think, is that instrument students become enamored with the ILS and practice it more than any other approach. As is the case with everything else in instrument flying, the more you practice something, the better you become at it.

The ILS is a precision approach because it incorporates an electronic glide slope. An ILS approach will bring you in exactly on the runway centerline if you fly the approach properly and it will take you down to within 200 feet of the runway when you break out at minimums. To do this, ILS provides very precise indications that you must respond to very precisely.

NEEDLE SENSITIVITY

By the time you reach the point in instrument training where you concentrate on ILS approaches, you probably will have practiced several without the hood. You know those needles are sensitive. The vertical needle is approximately four times more sensitive when set for the localizer of an ILS than it is on a VOR. And the horizontal glide slope needle is about four times more sensitive than the localizer indicator needle.

At the outer marker, a displacement of one dot equals approximately 300 feet on the localizer and 50 feet on the glide slope (FIG. 15-1). At the middle marker one dot equals 100 feet on the localizer and about eight feet on the glide slope. Only eight feet!

Localizer back course

Localizer transmitter:
1,000 feet from
opposite end of runway.

Glide slope transmitter

Middle marker (MM):
3,000 to 6,000 feet
from end of runway

Outer marker (OM)
Four to seven miles
from end of runway.

Localizer:
One dot = 100 feet
Glide Slope
One dot = 8 feet

Localizer:
One dot = 300 feet
Glide Slope:
One dot = 50 feet

Fig. 15-1. *Configuration of a standard ILS approach.*

Now more than ever you can begin to understand the importance of the standard of 2, 2, and 20—plus or minus two degrees on heading, two knots on airspeed, and 20 feet on altitude. If you have been working toward these goals throughout your instrument training, you should have no difficulty coping with the sensitivity of the ILS needles.

FLYING THE ILS

On an ILS final approach segment the basic instrument techniques must be very sharp. Overcontrolling will peg the needles and cause a missed approach.

To center the localizer needle, plan the turn onto the final approach course so that you roll out of the turn just as the needle centers. Quickly establish a reference heading that will correct for the wind, then use rudder pressure alone to make minor heading adjustments to the reference heading. Any bank at all will displace you from the localizer centerline so fast the needle will probably peg. Keep that localizer needle centered all the time. Avoid the temptation to make heading adjustments with bank.

When you begin the descent, set up a reference descent rate that will maintain the glide slope.

The next question is: What is the best descent rate for the approach? How do you determine what rate of descent will keep you on the glide slope? If you can find out, you will know what sort of power adjustment you need to make to set up that rate of descent.

A good method is to take the best estimate of the ground speed, divide by 2, and multiply by 10 for the rate of descent.

$$(80 \text{ knots} \div 2) \times 10 = 400 \text{ fpm}$$
$$(90 \text{ knots} \div 2) \times 10 = 450 \text{ fpm}$$
$$(100 \text{ knots} \div 2) \times 10 = 500 \text{ fpm}$$
$$(120 \text{ knots} \div 2) \times 10 = 600 \text{ fpm}$$

When you get the ATIS information for the landing runway, use this rule of thumb to estimate what the ground speed will be for the approach speed. If there is no ATIS and you can't estimate the ground speed accurately, use the indicated airspeed less 10 knots as the next best thing.

Then, as you begin descent on the ILS, reduce power 100 rpm (or 1″ manifold pressure) for each 100 feet rate of descent sought. If you estimate that your ground speed will be 80 knots, reduce power 400 rpm (or 4″ manifold pressure) to set up a 400 fpm descent.

It's interesting to note that if you are flying a Cessna 172 at an airspeed of 90 knots and you have to maintain a 600 fpm rate of descent to stay on the glide slope, that means the ground speed is 120 knots. You have a strong tailwind and if you have a short runway you might have to circle to land or you could run off the end of a short runway.

Once you set up a reference descent rate that more or less maintains the glide slope, leave the power alone. (Throttle jockeying is a form of overcontrolling.) Don't worry about the airspeed. Use elevator pressure alone to make minor pitch adjustments. "Pitch to the glide slope—power to the airspeed."

The glide slope needle becomes the "altimeter" for pitch. If you go above glide slope, use forward pressure to decrease pitch slightly and return to the glide slope. If you descend below glide slope use back pressure to establish level flight and re-intercept the glide slope. If you go below both the glide slope and the MDA, *execute an automatic missed approach immediately*. Obstacle clearance is not provided below MDA unless you are in a position to make a normal descent to a landing.

Remember how sensitive the glide slope needle is. You don't need to make a large correction to move eight feet in the vicinity of the middle marker!

If you are flying a retractable, intercepting the glide slope is even simpler: as you intercept, lower the gear. That will automatically produce the right descent rate to stay on the glide slope, with minor adjustments. It doesn't matter if the plane is a Mooney, Arrow, Aztec, Baron, Seneca, Aerostar, or an Aero Commander. Just drop the gear and that will set up a good rate of descent to stay on the glide slope.

ANALYZING AN ILS APPROACH

Let's turn now the ILS RWY 6 approach at Allentown-Bethlehem-Easton, Pennsylvania, (FIG. 15-2) and analyze it using the step-by-step process applied earlier to ADF and VOR approaches.

1. **Read the Fine Print.** In the lower left of the profile you will find the glide slope angle (GS 3.00) and the threshold crossing height (TCH 56). This information is provided for all ILS approaches. The threshold crossing height is the altitude in feet above ground level where the glide slope crosses the threshold.
2. **Check the Height of Obstacles.** Some tall ones within 10 nm of the airport (1,610 feet and 1,524 feet) and two rising above 500 feet in the vicinity of the final approach course.
3. **Pick the Correct Minimums for Airplane Category and Type of Approach.** Now the advantages of the ILS's greater precision become obvious. The altitude minimum for the straight-in approach to Runway 6 is only 594 feet. The visibility minimum is 24. This is a transmissometer-measured visibility of 2,400 feet (less than one-third the length of the landing runway).

DECISION HEIGHT

The 594-foot altitude minimum for the straight-in approach to Runway 6 is a decision height (DH), not a minimum decent altitude.

Decision height is the height at which a decision must be made during an ILS or other precision approach to continue the approach and land or to execute a missed approach. DH does not allow maneuvering that is possible with MDA. You cannot level off at DH and continue in the hope of seeing the field and landing. At DH you *must* decide to land or make a missed approach. These are the only options available at DH.

Fig. 15-2. *A typical ILS approach, the ILS RWY 6 at Allentown-Bethlehem-Easton, Pennsylvania.*

Even though DH is the minimum altitude on an ILS approach, you must also determine the MDA. If you drop below both DH and MDA on an ILS approach, a missed approach is mandatory. Obstacle clearance is not provided below MDA unless you are in a position to make a normal descent to a landing.

To continue the approach after reaching DH, you must comply with the criteria listed in FAR 91.116(c), "Operation Below DH or MDA." Summarized:

- The aircraft must continuously be in a position from which a descent to a landing can be made;
- The visibility is not less than that prescribed for the approach in use;
- One or more of the nine defined visual references must be distinctly visible and identifiable.

Note that there is an additional approach possible with an ILS: a localizer approach, shown as "S-LOC 6" at Allentown. A localizer approach utilizes the high precision localizer beam for course guidance, but has no glide slope information. If the glide slope transmitter at the airport goes off the air, or if there is a failure of the glide slope receiver in the airplane, you may continue the approach on the localizer alone. But the approach becomes nonprecision without the glide slope.

On a localizer approach, the minimum altitude is an MDA, not a DH. The circling minimums listed below the localizer minimums apply only to the localizer approach. You cannot circle to land out of a full ILS approach with glide slope, unless the descent is stopped at the circling minimums shown on the approach chart for that specific approach.

Check Adjustments for Inoperative Components Table. The inoperable components table is carried in the front of each NOS Instrument Approach Procedures booklet. See FIG. 14-2. The status of many more items is relevant to the minimums of an ILS approach. But it's not as complicated as it looks. Only a middle marker (MM) outage increases the DH, and the increase is 50 feet for all categories.

The visibility minimum increases with the outage of various approach and runway lights. The approach light code for Allentown Runway 6 is shown on the airport diagram (FIG. 15-3) at the approach end of the runway. Use the approach lights table on the inside back cover of the NOS Instrument Approach Procedures booklet (FIG. 14-3), to identify the type of lighting for the ILS landing runway. For Allentown Runway 6, the system is MALSR. If it goes out, the visibility minimum increases for Category A airplanes on the ILS approach. That certainly makes sense. If there is a failure of the lights, you will certainly need more visibility to find the runway, especially at night.

Be sure to examine higher minimums that might be required for inoperative components in the localizer approach. You will find the localizer increases lumped together in Section 3 of the table with many other nonprecision approaches.

Make Adjustments Required by Fine Print. None in this case.

Add the Altimeter Error. If you detect an altimeter error prior to flight, make no adjustments. Just *add* it to minimums on the approach.

Fig. 15-3. *LOC BC RWY 24, the back course approach at Allentown.*

Now run through the MARTHA check and make sure you understand all these elements as they apply to the intended approach, in this case the ILS RWY 6 approach at Allentown:

MA - Missed Approach. "Climb straight ahead to 3,000 via NE course of the localizer to NAZAR intersection and hold as shown." This is different. The procedure calls for you to track outbound on the back course of the localizer. When tracking a localizer back course outbound, the normal tracking procedure is used—correct toward the needle the same as the ILS front course (normal sensing outbound on an ILS back course).

R - Radios. Your No. 1 NAV will be set on the localizer frequency, 110.7. When you tune the localizer frequency, the glide slope is automatically received.

Note that the localizer frequency is underlined. That indicates no voice transmission capability. The identifier is I-ABE. All localizer identifiers have the prefix I to eliminate any confusion between localizers and VORs.

The No. 2 nav will be set for the radial that establishes the holding fix on the missed approach. The station is SBJ (Solberg VOR) on 112.9. You will dial the Solberg 304° radial with the OBS.

ADF will be set to the ILS Runway 6 compass locator, LEEHI, identifier AB.

MARKER BEACONS

Two marker beacons send up a very narrow VHF beam to fix an airplane's position on the ILS final approach course. Beacons are tuned automatically whenever the receiver is operating.

The outer marker transmits a continuous series of two audible dashes and a light flashes blue when you pass over it. (Students find it easy to remember the code if they think of it as saying "You're HERE, you're HERE, you're HERE.)"

The middle marker transmits a continuous series of audible alternating dots and dashes and an amber light flashes. Some ILS approaches—mainly at the larger and busier airports—also have an inner marker. The inner marker transmits a continuous series of dots and flashes white.

Now back to the MARTHA check:

T - Time. Pick the time from FAF to MAP based on the best estimate of ground speed. All ILS approaches should be timed. If the glide slope goes out you can continue with a localizer approach without resetting anything. Use the published MDA instead of DH and the MAP will be determined by timing the final approach segment from the FAF to the MAP.

H - Heading. The final approach course heading, in this case 061°.

A - Altitude. DH for the straight-in ILS 6 approach is 594 feet. MDA for the straight-in localizer 6 approach is 820 feet. MDA for circling approach out of the localizer approach is 840 feet.

ILS TIPS

Request the full procedure where available to get the most out of ILS training. Large, busy airports will probably turn you down because of the heavy flow of traffic. Search out an uncontrolled airport with an ILS where you can practice as many full approach procedures as you wish.

On VFR cross-countries (and at your home airport if it has an ILS) contact approach control and request "practice" ILS. Remain VFR and fly the approach unhooded when you don't have a safety pilot. (Don't forget collision avoidance—somebody must be looking!) Practice approaches will help you get the "big picture" of how ILS proceeds from step to step at different airports. In spite of common basic elements, all approaches—including ILS—are slightly different.

Practice holding on the localizer course (as shown on the Allentown ILS RWY 6 procedure, for example) without using the compass locator. It takes a little extra practice to set up a holding pattern on that very sensitive localizer needle. Turn outbound when the marker beacon starts to fade.

Always be prepared to switch from the full ILS to the localizer approach at any time.

Instructor Note. Some airplanes have circuit breakers that you can pull to simulate a glide slope failure. If you can't do this, simulate the failure by turning the receiver off. But be fair to the student. Turn it off early so he can retune a second receiver to the localizer frequency.

Be prepared to switch from an ILS or localizer approach to an NDB approach if there is a compass locator at the outer marker. Place the NDB approach chart beneath the ILS chart on the clipboard or yoke chart clip so you can look at it quickly if necessary.

Not all ILS approaches have compass locators or NDB approaches to the same runway. But if there is an NDB approach colocated with the ILS, it is excellent backup in case of transmitter or receiver failure.

Once you have tuned and identified the codes of the localizer and any VOR you might need, turn the volume down or the audio off. Failure in these two systems will cause warning flags to appear. On the other hand, adjust the volume on the ADF—after identifying it—to hear it faintly in the background. The only way you can tell if you have an ADF failure is by listening to the identifier. As long as you can hear it, all is well. (Unless you have inadvertently switched it to REC instead of ADF.)

Remember that a power reduction of 100 rpm (or 1″ of manifold pressure) produces a descent of 100 fpm; a power reduction of 500 rpm (or 5″ of manifold pressure) produces a 500 fpm descent at constant airspeed.

Avoid overcontrolling on the final approach course by using rudder pressure only— no banking—to keep the localizer needle centered. Use gentle elevator pressure to keep the glide slope needle centered. (Heading changes should be limited to two degrees, or at most five degrees, at any one time. Because the rule of thumb is "never bank more than one-half the degree of heading change,'' there is no way you can see a one-degree bank angle. So, why bother?)

''Pitch to the altitude'' and ''power to the airspeed'' on the glide slope.

ILS/LOC identifier signals are usually not clearly audible until you are at least within 40° of the final approach course. When you are ''abeam'' the transmitter site, all you hear is a lot of scratch, which might cause you to miss important communications.

BACK COURSE APPROACHES

ILS localizer antennas are located on the runway centerline about 1,000 feet beyond the far end of the approach runway. The localizer radiates both a ''front course,'' which is used for the ILS approach, and a ''back course'' (FIG. 15-1). The back course cannot be used for instrument approaches unless a specific approach procedure has been approved for that back course.

Allentown has a back course approach based on the ILS RWY 6 localizer: LOC BC RWY 24 (FIG. 15-3). Note the words BACK COURSE printed on the chart in large bold type. The reason for this warning is that back course approaches resemble conventional localizer front course approaches on approach charts. But they cannot be flown like front course approaches because the needle moves in the opposite direction, and no altitude (glide slope) information is available.

Reasons why the localizer needle moves in the opposite direction on a back course approach are fairly complicated and have to do with the way the localizer signal radiates from the antenna. It is more important to understand that when you make a back course approach, the needle moves *opposite* the way it does on a front course.

Always turn away from the needle to make a back course correction; turn to the left if the needle moves to the right; turn to the right if the needle moves to the left. This is the opposite of VOR or ILS corrections, where you *always* turn toward the needle.

Many students find it simpler to imagine that *they are the needle* and turn toward the bull's eye at the center of the instrument for correction.

A simple way to remember this is that when you are traveling in the same direction as the course for the normal ILS, you make corrections in the same direction as the needle. When you are traveling in the *opposite direction* from the normal ILS, as you do on a back course approach, you make *opposite corrections*.

One or two practice sessions with back course approaches will make all this clear. Some other points to consider when working with back course approaches:

1. Although they don't have glide slopes, the glide slope needle may come alive periodically. These are false indications. Ignore them.
2. The ''back course'' needle will be more sensitive than the front course because the localizer antenna array is usually located at the far end of the ILS front course runway (FIG. 15-1). Thus you will be operating just that much closer to it throughout the back course approach.

3. Some back course approaches have marker beacons at the FAF to indicate where the approach descent begins. These back course markers may be coded differently from those on the front course ILS. Back course markers transmit a continuous audible series of two dots and the white light flashes.

LOCALIZER, LDA, AND SDF APPROACHES

As we have seen, localizer approaches may be made on an ILS system whenever the glide slope is out. Many airports have localizers only with no glide slope equipment. This enables airports whose terrain and obstacle clearance rule out the glide slope required for the full ILS to have an approach with very precise centerline guidance.

FIGURE 15-4 shows the localizer approach at Pittsfield, Massachusetts. There are three features in this approach that we haven't encountered.

In the fine print below the profile: "Inoperative table does not apply." The minimums are already high because of surrounding mountains so inoperative component comments are not necessary.

The localizer (I-EIF, 108.3) also has DME. See the profile and note how DME is used as a cross-check at the IAF, FAF, marker, and MAP. DME is not required for the approach, however it would certainly be nice to have DME with high obstacles all around.

The third feature of interest is a fan marker rather than a marker beacon between the FAF and the MAP. This one is coded "R" as shown on the approach chart (dit-dah-dit) and activates the white light on the marker beacon panel.

Fan markers are similar to other marker beacons but more powerful. Fan markers have a power output of 100 watts, whereas ILS beacons have an output of three watts or less. Note the lower minimums (FM MINIMA) for a straight-in approach to Runway 26 with the fan marker.

LDA

LDA approaches are uncommon. LDA stands for Localizer-type Directional Aid. An LDA is nothing more than a conventional localizer that is not aligned with the runway, for instance the LDA RWY 2 approach at Hartford-Brainard, Connecticut (FIG. 15-5). The runway heading is 020°, but the localizer approach course is 001°, too great a divergence to qualify for approval as a localizer approach.

Straight-in approaches are allowed with an LDA when the divergence between the localizer course and the runway does not exceed 30°, as is the case at Hartford/Brainard. If the divergence is greater than 30°, only circling approaches may be made.

SDF

SDF stands for Simplified Directional Facility. See FIG. 15-6 for an example of an SDF approach. The SDF transmits a course similar to a localizer but not as precise. The beam of a true localizer varies between 3° and 6° to produce a width of 700 feet at the

Fig. 15-4. *A localizer approach at Pittsfield, Massachusetts, with two unusual features: DME with the localizer, and a fan marker on the final approach course.*

Fig. 15-5. *LDA RWY 2, a localizer directional aid approach at Hartford-Brainard, Connecticut.*

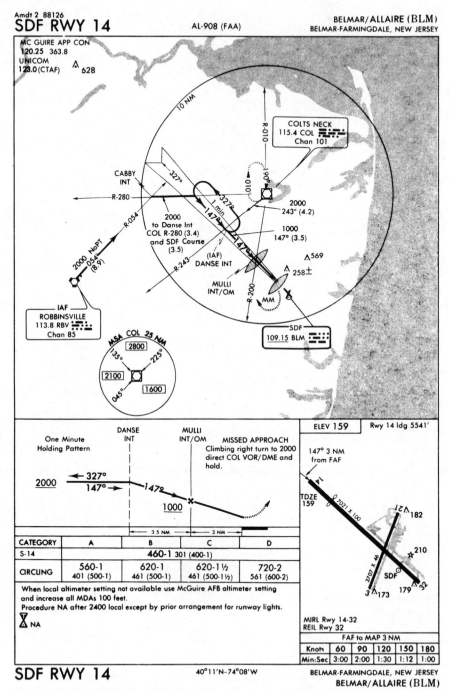

Amdt 2 88126

SDF RWY 14

AL-908 (FAA)

BELMAR/ALLAIRE (BLM)
BELMAR-FARMINGDALE, NEW JERSEY

MC GUIRE APP CON
120.25 363.8
UNICOM
123.0 (CTAF)

COLTS NECK
115.4 COL
Chan 101

CABBY INT

R-280

2000
to Danse Int
COL R-280 (3.4)
and SDF Course
(3.5)

(IAF)
DANSE INT

MULLI
INT/OM

MM

IAF
ROBBINSVILLE
113.8 RBV
Chan 85

2000
243° (4.2)

1000
147° (3.5)

569

258±

SDF
109.15 BLM

MSA COL 25 NM
2800
135° 225°
2100
045° 1600

ELEV 159   Rwy 14 ldg 5541'

One Minute
Holding Pattern

DANSE INT   MULLI INT/OM

2000   ← 327°
147° →   147°

1000

MISSED APPROACH
Climbing right turn to 2000
direct COL VOR/DME and
hold.

147° 3 NM
from FAF

TDZE
159

MIRL Rwy 14-32
REIL Rwy 32

CATEGORY	A	B	C	D
S-14		460-1 301 (400-1)		
CIRCLING	560-1 401 (500-1)	620-1 461 (500-1)	620-1½ 461 (500-1½)	720-2 561 (600-2)

When local altimeter setting not available use McGuire AFB altimeter setting
and increase all MDAs 100 feet.
Procedure NA after 2400 local except by prior arrangement for runway lights.

NA

	FAF to MAP 3 NM				
Knots	60	90	120	150	180
Min:Sec	3:00	2:00	1:30	1:12	1:00

SDF RWY 14

40°11'N–74°08'W

BELMAR-FARMINGDALE, NEW JERSEY
BELMAR/ALLAIRE (BLM)

Fig. 15-6. *SDF RWY 14, a simplified directional facility approach at Belmar/Allaire, New Jersey.*

landing threshold. However, the signal emitted by the SDF transmitter is fixed at either 6° or 12°. Think of the SDF sensitivity somewhere between a VOR radial and an ILS localizer. The SDF may also be offset from the runway centerline.

You do not need to demonstrate back course, localizer, LDA, or SDF approaches during the instrument flight test. But if there are any in your area, especially back courses, you should fly them whenever you have the opportunity. As a rated instrument pilot, you will be expected to execute these approaches whenever they are assigned by ATC. So practice now and avoid surprises and embarrassment later.

RADAR ASSISTS

These days you can expect radar assists on almost every approach. ATC monitors your en route progress with radar, and then hands you off to RAPCON (Radar Approach Control) who then vectors you to the final approach course for the procedure in use. Approach control will often turn you right onto the final approach course. (Procedure turns are prohibited.)

This doesn't mean that you shouldn't master the full procedures. On the contrary, you must know the full procedure for every approach you fly in order to visualize what approach control is requesting. Be prepared to go to the full procedure if you lose radio communications.

This raises an interesting question. Suppose you are receiving radar vectors to intercept the final approach course of an ILS. But you have not yet been cleared for the approach itself and you lose radio communications at that point. You just don't hear anything on either receiver or any voice frequency you might have tuned in. What do you do?

You carry out the lost communications procedure as specified by FAR 91.127. Summarized:

VFR. ''Continue the flight under VFR and land as soon as practicable.'' In other words, break off the instrument approach and enter the normal VFR traffic pattern.

IFR. Continue with the ''route assigned in the last ATC clearance received.'' In this case the ''route'' would be ''radar vectors to the final approach course.'' Turn to intercept the final approach course and complete your approach and land or make a missed approach and depart for the filed alternate.

Instructor Note. Check the full text of FAR 91.127 and teach it correctly. Remember, if the student hasn't learned the instructor hasn't taught! Don't assume lost communications just because approach control hasn't talked to you lately. They may have their hands full with an emergency or some other serious situation. Remind ATC: ''Cessna five six Xray, final approach course imminent, request further clearance,'' or whatever covers the situation, and they should respond.

Radar monitoring and radar vectors do not constitute radar approaches, however. To get a radar approach, you must normally request it. Radar approaches may also be offered to airplanes in distress or to expedite traffic.

ASR APPROACHES

The most common type of radar approach is the airport surveillance radar (ASR) approach, or surveillance approach. Look in the front section of any NOS Instrument Approach Procedures booklet to find a section that lists radar approaches available in the area covered by the booklet, along with their minimums (FIG. 15-7). Note the DH/MDA column. ASR approaches don't get very low.

Once approach control assigns an ASR approach, they will tell you *exactly* what to do. If you don't have the ATIS, they will provide it; they will provide lost communications and missed approach procedures; they will provide radar vectors; and they will provide runway alignment.

Approach control will issue advance notice of where descent will begin, and if the pilot requests it, approach control will also provide recommended altitudes on final approach. You will hear this kind of phraseology:

> *"Prepare to descend in _____ miles."*
> *"Published minimum descent altitude _____ feet."*
> *"_____ miles from runway. Descend to your minimum descent altitude."*

Approach control will keep talking to you on final and inform you if you are deviating from the final approach course:

> *"Heading _____, on course,"* or *"slightly (or) well left (or) right of course."*

As you get closer they will inform you of your distance from the MAP:

> *"_____ miles from missed approach point,"* and, if requested *"recommended altitude is _____."*

Surveillance approach guidance may be discontinued when the pilot reports the runway in sight. Approach will say:

> *"_____ miles from runway (or 'over missed approach point'), take over visually."*
> *"If unable to proceed visually, execute a missed approach."*

NO-GYRO APPROACHES

An airplane on partial panel can be given a "no-gyro approach." In this radar procedure, all turns are started and stopped by approach control. The pilot is expected to make standard rate turns until turning onto final approach, when all turns are half standard rate. Turns should be started immediately upon receiving instructions. The instructions

RADAR INSTRUMENT APPROACH MINIMUMS

ALBANY COUNTY NY   Amdt. 14, JUN 23, 1983   ELEV **285**
RADAR—125.0 307.2

	RWY	GS/TCH/RPI	CAT	DH/ MDA-VIS	HAT/ HAA	CEIL-VIS	CAT	DH/ MDA-VIS	HAT/ HAA	CEIL-VIS
ASR	1		ABC	**720**—¾	435	(500—¾)	D	**720**—1	435	(500—1)
	19		ABCD	**740**/50	460	(500—1)				
	28		AB	**780**—1	495	(500—1)	C	**780**—1¼	495	(500—1¼)
			D	**780**—1½	495	(500—1½)				
	10		AB	**1000**—1	723	(800—1)	C	**1000**—2	723	(800—2)
			D	**1000**—2¼	723	(800—2¼)				
CIRCLING			AB	**1000**—1	715	(800—1)	C	**1000**—2	715	(800—2)
			D	**1000**—2¼	715	(800—2¼)				

Categories A, B S—1 visibility increased ¼ mile for inoperative MALSR.
Inoperative table does not apply to MALSR Rwy 19 for Categories A and B.
Category C S—19 visibility increased ¼ mile for inoperative MALSR.
▼ ▲

BINGHAMTON/EDWIN A. LINK FIELD-BROOME   Amdt. 7, JUNE 5, 1986   ELEV **1630**
COUNTY NY
RADAR—118.6 397.2

	RWY	GS/TCH/RPI	CAT	DH/ MDA-VIS	HAT/ HAA	CEIL-VIS	CAT	DH/. MDA-VIS	HAT/ HAA	CEIL-VIS
ASR	34		AB	**2020**/24	423	(400—½)	C	**2020**/40	423	(400—¾)
			D	**2020**/50	423	(400—1)				
	16		ABC	**2000**—½	370	(400—½)	D	**2000**—1	370	(400—1)
	28		ABC	**1940**—1	353	(400—1)	D	**1940**—1¼	353	(400—1¼)
	10		ABC	**1960**—1	369	(400—1)	D	**1960**—1¼	369	(400—1¼)
CIRCLING			A	**2060**—1	430	(500—1)	B	**2080**—1	450	(500—1)
			C	**2080**—1½	450	(500—1½)	D	**2180**—2	550	(600—2)

Category D S—16 visibility increased ¼ mile for inoperative MALSR.
Category D S—34 visibility increased to RVR 6000 for inoperative SSALR.
▼ ▲

BUFFALO/GREATER BUFFALO INTL NY   Amdt. 13, APR 10, 1986   ELEV **724**
RADAR—123.8 236.5

	RWY	GS/TCH/RPI	CAT	DH/ MDA-VIS	HAT/ HAA	CEIL-VIS	CAT	DH/ MDA-VIS	HAT/ HAA	CEIL-VIS
ASR	5		ABC	**1100**/24	394	(400—½)	D	**1100**/50	394	(400—1)
	23		ABC	**1100**/24	376	(400—½)	D	**1100**/50	376	(400—1)
	14		ABC	**1100**—1	390	(400—1)	D	**1100**—1¼	390	(400—1¼)
	32		AB	**1120**—1	406	(400—1)	CD	**1120**—1¼	406	(400—1¼)
CIRCLING			AB	**1180**—1	456	(500—1)	C	**1180**—1½	456	(500—1½)
			D	**1280**—2	556	(600—2)				

Category D S—23 visibility increased to RVR 6000 for inoperative ALSF.
Category D S—5 visibility increased to RVR 6000 for inoperative SSALR.
▼

NE-2

RADAR INSTRUMENT APPROACH MINIMUMS

K1

Fig. 15-7. *Radar Instrument Approach Minimums listed in the front of every NOS Instrument Approach Procedures booklet.*

couldn't be easier to follow, consisting of such directions as "turn right," "stop turn," "turning final, make all turns one-half standard rate."

Radar controllers need to practice surveillance and no-gyro approaches, so make a point of requesting this service frequently enough to keep everyone proficient.

Instructor Note. ASR approaches may be available for airports other than those listed in the front section of the NOS Instrument Approach Procedures booklet. Telephone the RAPCON serving your area—or go visit the facility—and discuss your training needs with a supervisor. They can inform you which airports are best for ASR practice and which ones can't handle practice ASR approaches because of heavy traffic.

PAR APPROACHES

It's too bad PAR—precision approach radar—approaches aren't widely available. They are easy to learn, easy to use, extremely accurate, and there are no needles to keep centered. Ask any current or former military pilot about PAR—known in the military as GCA or ground controlled approach—and you will hear high praise for this precision approach.

With PAR, an electronic runway centerline and an electronic glide path are transmitted from equipment located alongside the runway in use. The precision radar tracks the incoming plane and shows it on a scope in relation to the electronic centerline and glide path. The radar is so sensitive that it can detect and display deviations of a few feet. Experienced controllers monitor the displays and tell incoming pilots what action to take to return to the centerline or to get back on the glide path.

A PAR approach begins like an ASR approach. The airplane is vectored onto the final approach course at a specified altitude. As it nears the electronic glide path, the final controller talks the pilot down. The sequence of instructions from the final controller runs something like this:

"Approaching glide path." (Ten to 30 seconds before final descent.)
"Begin descent." (On reaching the point where final descent is to start.)
"Heading _____. On glide path, on course." (To hold the airplane on course and on glide path.)
"Slightly above glide path, slightly left of course."
"Well above glide path, well left of course."
"Above glide path and coming down."
"Left of course and correcting."
"On course. On glide path."
"Three miles from touchdown."
"At decision height."
"Over approach lights."
"Over landing threshold. Contact tower after landing."

I don't know of any PAR approaches routinely available for civilian pilots to practice, nor did research for this book find one. So there is no opportunity to practice this approach.

However, the controllers are so skillful at "talking you down" that you won't need much practice if you ever have to use a PAR in an emergency. Just call the nearest military airport on 121.5, tell them your emergency, request a "GCA," and do what they tell you. They'll get you down safely in an expeditious manner.

VISUAL AND CONTACT APPROACHES

Two more types of instrument approaches exist that you should know about and be prepared to execute.

The first is the *visual approach*. As you arrive on an IFR flight plan, approach control may clear you for a visual approach to the airport or to follow another airplane. ATC cannot issue a visual clearance unless your approach and landing can be accomplished in VFR conditions. Approach control uses the visual approach to expedite incoming IFR traffic when the airport is VFR.

Some visual approaches are so common around big busy airports that approach charts have been developed. FIGURE 15-8 is the RIVER VISUAL RWY 18 approach to Washington National Airport, Washington, D.C. Approach control assigns visual approaches.

A *contact approach* must be specifically requested by the pilot; approach control cannot initiate a contact approach.

You may request a contact approach if you are flying an instrument approach and break out clear of clouds with at least one mile flight visibility and you can expect to continue to the destination in these conditions.

"One mile and clear of clouds" rings a bell, doesn't it? Right, these are the minimums for special VFR, and they are really *very marginal* minimums. Legal scud running, some would call it.

Never request a contact approach at a strange airport. Contact approaches should only be used at familiar airports. And they should only be when there is sufficient ceiling and visibility to depart from the instrument approach and enter a comfortable landing pattern.

INSTRUMENT TAKEOFFS

I like to introduce instrument takeoffs about midway in a student's training when he is able to fly the airplane by instruments in a very competent manner. This is usually the point where the student also begins concentrating on precision approaches, so instrument takeoffs are included in this chapter.

FAR 91.116(f) covers takeoff minimums under IFR. If you read that regulation carefully, you will see that it does not prescribe any IFR takeoff minimums for aircraft operating under Part 91. That's you and me. We may legally take off in any kind of weather. But if we do so when the ceiling and visibility are very low, we might be violating FAR 91.9, operating an aircraft in a "careless or reckless manner."

Amdt 3 88294

RIVER VISUAL RWY 18 AL-443 (FAA)

WASHINGTON NATIONAL (DCA)
WASHINGTON, DC

WASHINGTON APP CON
124.7 338.2
WASHINGTON TOWER
119.1 257.6
GND CON
121.7

CLNC DEL
128.25
ASR
ATIS 132.65

3000
Recommended Altitude

DCA 10 NM DME Arc

AMERICAN
LEGION
MEMORIAL
BRIDGE

DAVID W. TAYLOR
NAVAL SHIP RESEARCH
AND DEVELOPMENT CENTER

148°

1049

1049

1049

DALECARLIA
RESERVOIR

1800
Recommended Altitude

DCA 6 NM DME Arc

CHAIN
BRIDGE

P-56

GEORGETOWN
RESERVOIR

RIVER
REFERENCE
LIGHTS

1200
Recommended
Altitude

KEY
BRIDGE

P-56

900
Recommended
Altitude

DCA 4 NM DME Arc

DCA 3 NM DME Arc

ROOSEVELT
MEMORIAL
BRIDGE

ARLINGTON
MEMORIAL
BRIDGE

RADAR REQUIRED

Weather minimums: 3500
foot ceiling and 3 mile
visibility.

NOTE: Recommended Altitudes for
Noise Abatement

RIVER
REFERENCE
LIGHTS

GEORGE MASON
MEMORIAL BRIDGE

ROCHAMBEAU
MEMORIAL BRIDGE

WASHINGTON
111.0 DCA ⋅⋅⋅⋅
Chan 47

33

3 36

| 1 NM | 2 | 3 | 4 | 5 | 6 | 7 |

RIVER VISUAL RWY 18

Aircraft may visually follow the river to the airport, or may proceed via the DCA
VOR/DME R-328 (148° inbound) or via the Rosslyn LDA Rwy 18 Approach to
abeam Georgetown Reservoir or the DCA 4 NM DME fix, then follow the river to
the airport.
NOTE: Clearance for visual approach does not authorize penetration of P-56.

RIVER VISUAL RWY 18   38°51'N–77°02'W

WASHINGTON, DC
WASHINGTON NATIONAL (DCA)

Fig. 15-8. *The RIVER VISUAL RWY 18, a published visual approach to Washington
National Airport, as depicted on an NOS Instrument Approach Procedures chart.*

While it may be technically legal to take off when the ceiling and visibility are below IFR minimums, I think it is very poor judgment to do so. You should always be able to return immediately to the departure airport for an instrument approach if a problem develops. Furthermore, when you expect to climb into actual IFR soon after take off, you need a ceiling of approximately 200 feet to get established before entering the clouds.

However, practicing an instrument takeoff with a hood on is an exciting and interesting exercise. (It's always a revelation to students that they can do this, and their confidence grows considerably after they have tried a few.)

The procedure is simple. When cleared for takeoff, taxi out and line up on the centerline as usual. Hold the brakes and add full power. Release the brakes when you have three-quarters to full power and anticipate the tendency of the airplane to turn left during the roll by applying right rudder pressure.

Use rudder pressure to keep the airplane rolling straight down the centerline by reference to the heading indicator. Don't try to force the plane into the air. Let the airspeed build up five knots or so beyond normal lift-off airspeed, then apply back pressure. Pitch up to the first mark above the horizon on the attitude indicator and hold that attitude until a steady, positive rate of climb shows on the VSI and you have reached 500 feet above the airport. Then adjust the attitude for anormal climb.

You have to make corrections promptly to counter the tendency of the airplane to drift left during the takeoff roll. Fortunately, the heading indicator is sensitive enough so that you can detect very slight changes of heading. React quickly and positively to these slight movements of the heading indicator with rudder pressure.

With a little experience you will find that you can glance out the left window and see whether you are drifting left or right. This comes after you have done two or three instrument takeoffs and are confident enough to take your eyes off the heading indicator for a few seconds then look back to it right away without being distracted.

16
Putting it All Together: The Long IFR Cross-Country

THE LONG CROSS-COUNTRY IS THE CULMINATION OF EFFORTS AS A PREGRADUATION review of the real world of IFR. It meets the requirements of FAR 61.65 (c)(4), which calls for "cross-country flying, in simulated or actual IFR conditions . . . including one such trip of at least 250 nautical miles, including VOR, ADF, and ILS approaches at different airports." When you appear for the instrument flight test, one of the first things the examiner will do is check your logbook to make sure you have accomplished this long IFR cross-country as required.

I believe that if this long cross-country is attempted just to meet the bare bones requirements of the FAR, you're missing the true benefit and purpose of the flight. You can meet the qualifications by covering the distance and stumbling through the three different approaches, but you'll shortchange yourself.

This is the point at which the instrument student demonstrates to his instructor and to himself that he can do all the planning, fly from one place to another using the ATC system effectively, and make the different approaches down to minimums. Then all the instructor has to do after this lesson is review and polish up those elements that were found a bit lacking on the long cross-country. The student should then be ready for the flight test.

THE VALUE OF ACTUAL IFR

It doesn't make too much difference whether you have an actual IFR day or whether you have to simulate IFR with a hood. To me the ideal long cross-country has at least one leg of fairly heavy actual IFR. If the entire flight has to be completed in actual IFR conditions, great! It certainly builds confidence in the student to know that he has experienced actual IFR under the real conditions of a cross-country flight.

It is almost criminal to find that a number of instrument rated pilots have never seen the inside of a cloud. They file IFR and everything goes just fine until they see a bank of clouds ahead of them. Some will cancel IFR and duck under the clouds, which is just the opposite of what they were trained to do with complete mastery.

A person really hasn't earned an instrument rating unless he has experienced some actual IFR conditions, not merely flown through a few clouds. Actual IFR builds confidence in the student and shows him that it's not just make believe or theory during all those hours under the hood.

Once the initial shock wears off, it's much easier to fly IFR in actual conditions than to simulate it with a hood. The hood is restrictive and uncomfortable, and if you can legitimately get rid of it, the flight will seem more normal. Also, under actual IFR, the skies are certainly less crowded.

The long IFR cross-country required by the FAR is a basic, straightforward flight. If you have followed the syllabus carefully, you will have planned, filed and departed on many IFR cross-countries already.

UNCONTROLLED AIRPORTS

File three flights at the outset to terminate with the three different types of approaches required. Ideally, one approach should be to an uncontrolled airport. (One of the most common errors I see on instrument flight tests is a candidate making an approach at an uncontrolled airport without making traffic advisories on unicom (the CTAF). The candidate might invade the territory of other pilots and come barging into their air space unannounced, then leave without so much as a hello, thank you, or good-bye. It certainly gets the locals perturbed if they find themselves sharing the final with some clod who has a hood on! Or worse yet, who sets up a collision course with the traffic using the opposite runway.

One approach to an uncontrolled field will bring out any deficiencies in uncontrolled airport procedures. Most examiners make it a point to evaluate the way an instrument candidate conducts IFR approaches at an uncontrolled airport.)

If you haven't had much practice picking up a new clearance in the air, the long IFR cross-country is a good opportunity to do so. A person should not have an instrument rating attached to his pilot certificate without having at least some acquaintance with air filing and picking up a clearance in the air.

Plan to execute a missed approach, which will keep the IFR flight plan open. When ATC requests intentions before commencing the approach, tell them you have an IFR flight plan on file to your next destination. Be sure to have that flight plan handy so you can read it to ATC in case it has gone astray in the computer.

It should be pointed out that controllers are usually very considerate of pilots in the air when reading a clearance. They'll read it piecemeal and try not to issue more than two or three parts of a clearance at one time.

VOID TIME CLEARANCES

The long IFR cross-country also provides a good opportunity to practice void time clearances over the telephone when departing from an uncontrolled airport. Your flight plan has already been filed, so you will telephone flight service and ask them for the clearance. They may ask you to call back at a certain time for the IFR clearance, release, and void time. Void time means exactly that—the clearance is void after that time. If you see that you can't take off before the void time, you will have to call flight service again and request a more practical release time.

Void time clearances take a bit of forethought. You have to complete the preflight and get the charts and radios all set for departure before calling flight service for the clearance and release time. Position the airplane for a prompt departure.

I also like to do some partial panel work during a leg that is long enough so the student is well caught up on his cockpit workload. One thing I like to see on partial panel is the student tuning the ADF to an NDB or commercial broadcast station up ahead that will serve as a backup heading indicator, giving a constant relative bearing in combination with the magnetic compass.

As you plan the flight, ask what you would do if given a simulated vacuum system failure at various points along the way, then note what NDBs or commercial broadcast stations can function as a backup for the heading indicator at these different points. The examiner will certainly be impressed if you demonstrate this on the instrument flight test.

The fine points of IFR cross-country flying are well covered by Background Briefing 16-17 that precedes this flight lesson. However, there are two other aspects of cross-country flying that I would like to discuss in greater detail: fuel management and lost communications.

FUEL MANAGEMENT

Year after year, running out of fuel continues to be a major cause of general aviation accidents.

There is no excuse for this. Everyone knows, or should know, how much fuel is in the tanks at the beginning of the flight, the gallons or pounds per hour that will be consumed at various power settings, and how long it will take to complete the flight. Fuel requirements must be computed for every cross-country flight. The FARs require this. Fuel is time in your tanks. When will your fuel be exhausted?

What is the problem? There are really two problems.

First, many pilots rely too heavily on fuel gauges. The fuel gauges in general aviation light planes are not reliable, especially below one-quarter of a tank when the fuel is sloshing and surging around.

Second, you must break the habit of thinking in terms of "full" tank, "half" tank, "quarter of a tank," etc. Instead, think of hours and minutes of fuel remaining.

To do this, it is essential that you keep track of the time when tanks are turned on and off. Keep an accurate log of takeoff time and arrival time at all fixes and checkpoints. This isn't very hard because you have been logging these items on all cross-country flights, VFR as well as IFR, since student pilot days.

LOGGING THE FLIGHT

See FIG 16-1. It shows the filled-out log of the flight to Binghamton, New York, that was previously planned. It contains all the information that was entered during the flight.

Note the block for entering takeoff time (A), and the blocks for logging the times when tanks are normally switched (B). There was no need to switch tanks because we flew an airplane that feeds from both tanks.

The two columns under (C) provide blocks in which to log actual times en route on each leg and actual times of arrival at each fix. No actual times were logged for the flight to the alternate because the weather at Binghamton was VFR by the arrival time. See landing ATIS "Whiskey" at (D). (Above this is ATIS information "Bravo" for the departure airport, Westchester County.)

While looking at the log, note how the clearance was copied at (E), and how frequency and altitude changes were recorded at (F) and (G).

Also in block (A), the time of arrival was logged and the total time en route was computed and entered. To compute fuel remaining, subtract total time en route (1:10) from the hours and minutes in the tank on departure (5:30); (5:30) − (1:10) = 4:20. Think hours and minutes, not full and half tanks!

LOST RADIO CONTACT

Complete loss of communications while airborne on an IFR flight is extremely rare. What is more likely to happen is a loss of radio contact.

Many times lost radio contact is self-induced. The squelch may be tuned down too low or a wrong frequency may have inadvertently tuned. Double check the settings and audio panel. Also try transmitting on the other radio. It is possible for one to fail and the other to work just fine.

You may be flying through a quiet zone along the airway, or flying low enough so that the VHF line-of-sight transmissions are blocked by mountains or other obstructions.

Here is typical scenario: ARTCC to ARTCC hand off. Let's say Boston Center calls you with a transmission: *"Cessna five six Xray, contact New York Center, one two eight point five."*

You acknowledge the hand off by reading back the clearance: "New York Center one two eight point five, five six Xray," then you tune the new frequency and report, using the full N Number because this is an initial contact, and you report your altitude: "New York Center, Cessna three four five six Xray at niner thousand."

No response. Suddenly all is quiet.

DATE **November 30**   PILOT: **Sollman/Harris**

FLIGHT PLAN ROUTE

Westchester 9 Haays △
D-N40 V252 Lahy △

ATC CLEARS N 3456 X
C – BGM W-9 Haays F
Ex 90/0 120.55 1761

TAS: 125	IAS:	MP: 23'

RPM: 2300   "B" 7⊕ 2F

TAKEOFF:
RWY 34   WIND 300   KTS 5   ALT 0.04

Field Elev: 439
ALTIMETER ERROR: **40**

LANDING: **G**-W: 120 ⊕ 6
Field Elev: 630
RWY 34   WIND 300   KTS 19   ALT 0.08

TYPE OF APPROACH: ILS 34

TIME ON: **A** 1505
TIME OFF:
TOTAL TIME ENROUTE: 70 (1+10)

TACH ON 1235.9   TACH OFF 1236.8

CALL SIGN	FREQ	ASGND ALT
ATIS	133.8	
CPT	127.25	30
GRND	124.8	90
TWR	119.7	80
DC	120.55	40
F BOS	120.2	
NY	128.5	
ATIS	128.15	
AC	118.0	
TWR	114.3	
GRND	121.9	

CLEARANCE LIMIT: BGM

WINDS ALOFT

Sta	A	Dir	Vel	Temp	Dr	A			V			T
JFK	3	18	24	-	14	28	+8	9	20	32		+1
AVP	23	6	-		20	27	+7	20	42			+1

VFR WX AT AVP (Wilkes-Barre)
0900 - 20© @ 25 49/42 28⊕/13

TANK ON 1505   TANK OFF
B

Flight log table

FIX NAME	FIX FREQ / IDENT	INTERSEC NAME / IDENT / FREQ / RADIAL	ROUTE VIA / ALT	M.C. TO / FROM	DIST TO / REM	G/S EST / ACT	TIME ENR EST / ACT	TIME ARR EST / ACT
–		Haays STW 109.6 054	W-9	305 / 325	31+	108	26 / 31	EST 31
N40 116.1				325		144	28 / 33	33
Huguenot			V273	325 / 34	8		4 / 37	37
N40 116.1							5 / 38	38
		Lahy ANK 116.8 272	V252	38 / 66		134	30 / 08	08
BGM 112.2							28 / 06	06
BGM - airport			ILS 340	14	14	14	8 / 14	14
I-BGM 110.3							9 / 15	15
					125	68	(1+08)	
(Alternate)								
Lake Henry			V114 155	50	127	28		
LHY 110.8				301				
		Cosey IZK 109.4 035°	D 301	4	82	3		
LHY 110.8			ILS 225	13	100	8		
AVP - airport					67	0+39		
I-IBt 109.9				+125		1+08		
				192	1+47			
				Reserve 45	2+32			

Bottom label row: AC ID | POSITION | TIME | ALT | IFR/VFR | EST NEXT FIX | NEXT FIX | PIREPS

Fig. 16-1. *The completed flight log to Binghamton showing how items were entered during the flight. Letters indicate general sequence in which the items begin filling in.*

You try two or three more times and still get no response. What do you do? Make sure you haven't made some mistake like switching from speaker to headphones.

Return to the frequency you just left and report "Cessna five six Xray, unable New York Center one two eight point five." Boston Center should answer. They might tell you to remain with them, or attempt to contact New York Center on another frequency, or to stay with Boston and try New York again in a few minutes or a few miles.

Now suppose you had returned to the Boston Center frequency and couldn't raise them again. Nothing doing on either New York ahead or Boston behind you. Now what?

1. Check for other center frequencies in the area (FIG. 16-2). In this case, the only other frequency available is a UHF frequency, 363.2, which you don't have aboard.
2. Attempt to raise another aircraft to relay the message. If you hear one, give him a call. If not, transmit in the blind: "Any aircraft, Cessna three four five six Xray, requesting a communications relay." If someone answers, have him try to reach the new center—New York in this case—and relay their instructions.
3. Try contacting the nearest approach control or even a nearby tower and ask them to relay your message to the center.
4. Call flight service on 122.1, and give them a nearby VOR frequency with voice capability that you can hear clearly. Be patient. It may take them a couple of minutes to complete another transmission before they can get to you.
5. Come up on the emergency frequency 121.5, say you're having communications difficulties, and if anyone reads you, please reply.

(Many pilots are intimidated by the emergency frequency. They are reluctant to use it because they think they will have to write a lengthy letter to the administrator in Washington justifying the action. This is usually not the case. You don't have to report to anyone just because you used the emergency frequency.

Fig. 16-2. *Box containing sector frequencies for an Air Route Traffic Control Center (ARTCC).*

The emergency frequency is intended for lost radio contact as well as for more serious situations. When all else fails, try to re-establish communications on this frequency.)

6. When you switch to 121.5 you also squawk 7700 on the transponder. As you know, this is the emergency transponder code. It shows up on radar as a flashing digital display. You will attract the attention of everyone looking at an air control radar covering your area—military as well as civilian—when you squawk 7700. To indicate that you have a communications emergency, squawk 7700 for one minute, then switch to 7600 for 15 minutes, then back to 7700 for one minute and so on, alternating these 16-minute cycles for the remainder of the flight.

7. Try to repair the radio. You may have a hot microphone or a mike button stuck in the keyed position. You can find this out very quickly by unplugging the mike from its jack. If you start hearing other transmissions again, the mike is the problem.

A muting relay engages when you depress the microphone key. It cuts out incoming signals when you transmit, because you can't talk and listen at the same time without getting a howling, screeching feedback that will blast your eardrums out of action. So if the mike button is stuck, the muting relay will be engaged and you won't hear anything.

Because a stuck mike transmits continuously it also blocks everything on that frequency. When you unplug the stuck mike, you will begin hearing normal transmissions again, plus a lot of unkind remarks about the idiot with the stuck mike. So help everyone out when you have a radio problem and check for a stuck mike. If it sounds like you are still transmitting after you take your thumb off the mike button, check for a stuck mike first before trying any of the other alternatives.

You might also try giving the transceivers a strong shove back into their racks. Sometimes they come loose and disengage the connections at the back.

TWO-WAY RADIO COMMUNICATIONS FAILURE

If you have tried every method of restoring radio communications listed above and nothing works, then you must assume a two-way radio communications failure. FAR 91.127 (Appendix B) spells out very clearly the procedure you must follow when this happens.

IMPORTANCE OF LOGGING TIMES

Note how important it is in (i) and (ii) to know the estimated time of arrival (ETA) at your destination if you have not received an "expect further clearance time." ATC will expect you to show up for an approach at an estimated time of arrival (ETA) based on your filed or amended estimated time en route (ETE).

ATC knows when you took off from the departure airport (do you?), and they know your estimated time en route from your filed or amended flight plan. They will add the time en route to your departure time and come up with an ETA. ATC will reserve all the approaches for you at the destination for this ETA plus 30 minutes. Until that time expires, you "own" that airport.

So it becomes extremely important to know what time you took off, because without that information, you'll never know when to commence the approach. ("You can't tell when you're going to get someplace if you don't know when you left someplace.")

This seems so basic, yet many pilots forget to log their takeoff time. When I notice this on an instrument flight test, you can bet that at some point I will ask that would-be instrument pilot to tell me what his destination ETA will be in the event of lost communications or to detail his actions in case of communications failure.

EMERGENCY ALTITUDES

A second point that always raises questions is the altitude for completing the flight in the event of a two-way radio communications failure.

ATC expects you to continue at (1) the last assigned altitude, *or* (2) the minimum altitude for IFR operations, *or* (3) the altitude that ATC has advised you to expect in a further clearance. You will fly at the highest altitude of the three choices for any given leg.

You will have to make some choices. These choices hinge on how the "minimum altitude for IFR operations" is determined, and whether or not it is higher than the altitude assigned or advised to expect.

Along an airway, the minimum for IFR operations is the MEA or minimum en route altitude. If the En Route Low Altitude Chart shows that you are approaching a route segment with an MEA *higher* than that assigned by ATC or advised to expect, you would begin a climb to reach the new altitude so that you reach it prior to the point or fix where the new altitude begins. If the MEA drops down below the last assigned altitude, you would descend to the last altitude assigned by ATC or one that you were advised to expect, whichever is higher.

But suppose you were cleared direct. What is your minimum altitude off airways where there are no MEAs?

Consult the appropriate sectional chart that covers the flight area and pick out the "maximum elevation figure" for the latitude-longitude square you occupy, then add 1,000 feet to comply with FAR minimum altitude regulations (2,000 feet additional in mountainous terrain). The maximum elevation figure is the big numbers in the center of each latitude-longitude square. You would use the "maximum elevation plus a thousand" figure as a substitute for the MEA.

The likelihood of having to use the two-way radio communications failure procedures is remote, but you have to know them. There is a task listed in the *Instrument Rating Practical Test Standards* that says: "Exhibits adequate knowledge of two-way radio communications failure procedures."

In the real world of IFR you can avoid all the complications of lost communications by investing in a hand held portable transceiver and carrying it on every flight. It will also be invaluable in the next scenario.

COMPLETE ELECTRICAL FAILURE

Single-engine light planes are vulnerable to complete electrical failure. There is usually only one alternator—driven by one inexpensive V-belt—and only one voltage regulator. So there is no backup if one of these key components fails.

The first step in the event of a complete electrical failure is to turn off all electrical equipment to minimize drain on the battery. Control the airplane by partial panel. At night you will need a flashlight to do this. Careful pilots carry two flashlights for this purpose, one as a backup in case the other starts to fade.

If the failure occurs when you are VFR, maintain VFR and land as soon as practicable, just as you would if you had two-way radio communications failure.

If you are in actual IFR conditions when you have a complete electrical failure, the problem is equally simple, although your work is cut out for you. Find VFR conditions and land as quickly as possible.

When planning for the flight you filled in the block "VFR WX AT:" (FIG. 16-3); you know where you want to go. If you have been logging the actual times of arrival at each fix along the way, you have an accurate idea about where you were when the failure occurred. Draw a straight line to the new destination from the point you intend to turn. Calculate a new magnetic heading and turn to that heading. Later, with more time, calculate an ETA at the new destination. It's like those "diversions" you worked on so hard during the cross-country phase of private pilot training.

Fig. 16-3. *Flight log showing filled-in block for nearest VFR weather.*

When everything is under control, set the transponder to 7700 and turn it on. Then turn on the transceiver on which you last talked to ATC. You have only a few minutes of battery power left; make sure ATC understands your predicament and intentions, then shut everything off. You can put the remaining battery power to good use at the VFR destination to contact a tower or show some lights.

Again, if you have a portable transceiver aboard, it will make life a lot easier. Range of these units is vastly improved if connected to an aircraft antenna; consider installation of an antenna jack.

IFR CROSS-COUNTRY TIPS

Avoid marking up the chart with a highlighter. This can add to confusion on subsequent flights in the same area. I make it a game to see how much of an IFR flight I can complete without consulting the Low Altitude En route Charts except for an amended clearance. A previously marked up chart can be hazardous, especially if you have made a series of flights over the same area since the last chart revision.

After planning the flight, talk yourself through it. Make sure you understand where all those obscure intersections are, and note the airports along the way, especially those with instrument approaches.

Carry VFR charts to cover the route. Highlight the IFR route on the VFR charts for quick reference in case VFR landmarks are needed.

Save the flight logs. You might frequently use the same routes, or portions of them, over and over again. ATC often has preferred routings; old logs will highlight the preferences and you can plan for them and avoid the frustration of filing for one route and getting another.

Visit the center and the RAPCON that handles your local area. Call ahead for an appointment; you will find the supervisors and controllers most cooperative in showing you around and answering questions, especially if you can make an appointment for one of their less busy times.

Instructor Note. ATC centers and RAPCONS prefer to brief groups of 10 to 25 people. Visit these facilities in person and find out who can set up a briefing and what are the best dates and times. Post a sign-up sheet in the flight school or FBO and get a group together. Invite everyone. Groups are always more interesting because everyone has different questions and different experiences to bring up. Some of the questions will surprise you!

17
How I Conduct an Instrument Flight Test

WHEN A CANDIDATE SCHEDULES AN INSTRUMENT FLIGHT TEST, THERE ARE A NUMBER of things that I, as a designated examiner appointed by the FAA, must check before the test, as all examiners must do.

REQUIRED DOCUMENTS

First of all, is the airplane legal to make this flight? You must bring engine and airframe logs for the airplane to be flown and you must show the entries that indicate the 100-hour (if required) and annual inspections have been done within their deadlines, and that the pitot static and transponder systems have been tested within the last 24 months.

You must be able to open the logbooks to the pages with the entries, show them to me, and be familiar with airworthiness directives for the airplane and show that they have all been complied with. Find the entries ahead of time and mark the pages with paper clips to save time on the day of the test (and it will make a good first impression on the examiner).

Better yet, make a list of all these airplane requirements along with the hours and/or dates and be able to find the supporting information in the aircraft logs.

Be prepared to show the airplane's current weight and balance data, and an operating manual if one is required to be carried on board for that airplane. Prepare a weight and balance calculation for the flight test.

The next question is, are you legally eligible to make this flight?

Fig. 17-1. *Checklist for instrument flight test, based on* Instrument Rating Practical Test Standards.

ACCEPTABLE AIRCRAFT:

- ☐ View-limiting device
- ☐ Aircraft documents:
  - ☐ Airworthiness certificate
  - ☐ Registration certificate
  - ☐ Operating limitations
  - ☐ Weight and balance form
- ☐ Airframe and engine logs
- ☐ FCC station license

PERSONAL EQUIPMENT:

- ☐ Current aeronautical charts
- ☐ Computer and plotter
- ☐ Flight plan form
- ☐ Flight logs
- ☐ Current AIM

PERSONAL RECORDS:

- ☐ Pilot Certificate
- ☐ Medical Certificate
- ☐ Completed Airman Certificate and/or Rating Application (FAA Form 8710-1)
- ☐ Airman Written Test Report (AC Form 8080-2)
- ☐ Logbook with instructor's endorsements
- ☐ Notice of Disapproval (if applicable)
- ☐ Approved school graduation certificate (if applicable)
- ☐ Examiner's fee

REQUIREMENTS FOR INSTRUMENT RATING

I will review your application, FAA Form 8710-1, (FIG. 17-2) and pilot logbook to make sure you have satisfied all the requirements specified by the FARs for this flight.

To be eligible for the instrument rating, FAR 61.65 requires:

1. 125 hours of pilot time, of which 50 hours are as pilot-in-command in cross-country with other than a student certificate. To qualify as a cross-country, the flight must have a landing more than 50 nautical miles from the original departure point.
2. 40 hours of simulated or actual instrument time, of which not more than 20 hours may be in an instrument ground trainer acceptable to the FAA administrator.
3. 15 hours of instrument flight instruction by an authorized flight instructor, including at least 5 hours in an airplane or a helicopter, as appropriate.

TYPE OR PRINT ALL ENTRIES IN INK

Form Approved OMB No: 2120-0021

Airman Certificate and/or Rating Application

U.S. Department of Transportation
Federal Aviation Administration

I Application Information

☐ Student
☐ Private
☐ Commercial
☒ Instrument
☐ Other _____

☐ Additional Aircraft Rating
☐ Flight Instructor
☐ Ground Instructor
☒ Airplane Single-Engine
☐ Airplane Multiengine

☐ Helicopter
☐ Glider
☐ Lighter-Than-Air
☐ Additional Instructor Rating
☐ Medical Flight Test

☐ Reexamination
☐ Reissuance of _____ Certificate
☐ Renewal of _____ Certificate
☐ Renewal of Flight Instructor Certificate
☐ Reinstatement of Flight Instructor Certificate

A. Name (First, Middle, Last)	B. SSN	C. Date of Birth Mo. Day Year	D. Place of Birth
JOHN J. DOE	249-51-4529	02-31-56	NEW YORK, NY

E. Address	F. Nationality Specify		G. Do you read, speak and understand English?
1010 EAST 86th STREET	☒ USA ☐ Other_____		☒ Yes ☐ No

City, State, Zip Code	H. Height	I. Weight	J. Hair	K. Eyes	L. Sex
NEW YORK, NY 10021	71 In.	175 Lbs.	BROWN	BROWN	☒ Male ☐ Female

M. Do you now hold, or have your ever held An FAA Pilot Certificate? ☒ Yes ☐ No	If yes, has certificate ever been Suspended or Revoked ☐ Yes _____ Date ☒ No	N. Grade PVT	O. Certificate Number 249514529	P. Date Issued 01-08-88

Q. Do you hold a Medical Certificate? ☒ Yes ☐ No	R. Class of Certificate THIRD	S. Date Issued 04-31-88	T. Name of Examiner RICHARD R. ROE, MD

U. Have you ever been convicted for violation of any Federal or State statutes pertaining to narcotic drugs, Marijuana, and depressant or stimulant drugs or substances? ☒ No ☐ Yes	V. Date of Final Conviction

Glider or Free Balloon Pilots only:	Medical Statement: I have no known physical defect which makes me unable to pilot a glider or free balloon	W. Signature	X. Date

II Certificate or Rating Applied For on Basis of:

☒ A. Completion of Required Test	1. Aircraft to be used (if flight test required) CESSNA 172	2a Total time in this aircraft 50 hours	2b Pilot in command 45 hours

☐ B. Military Competence Obtained in	1 Service	2 Date Rated	3 Rank or Grade and Service Number
	4 Has flown at least 10 hours as pilot in command during the past 12 months in the following military aircraft.		

☐ C. Graduate of Approved Course	1 Name and Location of Training Agency		2 Agency School Number
	3 Curriculum From Which Graduated		4 Date

☐ D. Holder of Foreign License Issued By	1 Country	2 Grade of License	3 Number
	4 Ratings		

III Record of Pilot time (Do not write in the shaded areas.)

	Total	Instruction Rec'd	Solo	Pilot in Command	Second in Command	Cross Country Instruction Received	Cross Country Solo	Cross Country Pilot in Command	Instrument	Night Instr. Rec'd	Night Take-off/ Landing	Night Pilot in Command	Night Takeoff/ Landing Pilot in Command	Number of Flights	Number of Aero-Tows	Number of Ground Launches	Number of Powered Launches	Number of Free Flights
Airplane	126	60	20	90		12	10	50	41	7	20							
Rotor-Craft																		
Gliders																		
Lighter than Air																		
Instrument Ground Trainer	8	8							8									

IV Have you failed a test for this certificate or rating within the past 30 days? ☐ Yes ☐ No

V Applicant's Certification I certify that the statements made by me on this application are true.	A. Signature John J. Doe	B. Date 02-14-89

FAA Use Only

EMP	REG	D.O.	SEAL	CON	ISS	ACT	LEV	TR	S.H.	SRCH	#RTE		RATING (1)

FAA Form 8710-1 (6-86) Supersedes Previous Edition

Fig. 17-2. *Completed application for instrument flight test, FAA Form 8710-1.*

FAR 61.65 also requires "cross-country flying, in simulated or actual IFR conditions, on Federal airways or as routed by ATC, including one such trip of at least 250 nautical miles, including VOR, ADF, and ILS approaches at different airports."

Show me your current medical certificate and private or commercial pilot certificate. When you pass the flight test you must surrender the pilot certificate. A temporary certificate will be issued that includes the new instrument rating; you will receive the new permanent certificate by mail within 120 days.

In addition, you show me the entry for a current biennial flight review or equivalent. Again, put a paperclip on the logbook pages that contain the appropriate entries and you will save time.

I must examine your logbook entries for other items. For example, if you mistakenly included the cross-country time you flew as a student pilot while preparing for the private pilot certificate, you may not have enough time to meet the requirement of 50 hours other-than-student cross-country-flying for the instrument rating.

Sometimes I find people with an entire flight logged as instrument time. The total flight time was two hours and they logged two hours of simulated instrument time. No, total time is "block-to-block" or Hobbs time. Even if it was low ceiling IFR that day, they could not log the whole flight as instrument time. Some of this time had to be spent getting clearances, taxiing out, getting off the ground, landing, and taxiing back. The maximum instrument time you might get on a two-hour instrument flight is 1.75 hours.

I also must verify that the long IFR cross-country covered 250 nautical miles and the required approaches. The entry for that flight should indicate where you went and which instrument approaches you made and at which airports.

ENDORSEMENTS

The next items in the pilot's logbook are the two required endorsements. The first certifies that your flight instructor has given you the ground instruction required by FAR 61.65 (b)(1) through (4). The second certifies that you have received the flight instruction specified by FAR 61.65 (c)(1) through (5). The recommended form for these endorsements is shown in Chapter 2.

If this is a retest, the flight instructor must also enter another endorsement to the effect that he has given additional flight instruction required to cover the deficiencies found on the previous flight test.

On a re-test, you are expected to comply with all other requirements, just as though this was the original test, including a newly-completed application (FAA Form 8710-1). You must also present the notice of disapproval from the first flight—the pink slip.

The reverse side of FAA Form 8710-1 must also be signed by your instructor. And the instructor must endorse the written test report in the space provided to show that he has provided instruction on the items, if any, you missed on the written test. The FAA wants you to be 100 percent in all respects to earn your rating.

I find it helpful to instrument students and instructors alike to have a checklist they can review when the instructor is ready to recommend a student for a flight test (FIG. 17-1). It is based on the checklist in *Instrument Rating Practical Test Standards*. The checklist in FIG. 17-1 also contains some routine items not covered in the discussion above.

Portions of the application (FAA Form 8710-1) should be clarified. The application should be typed using capital letters. Dates should be written with two digits for date, month, and year. For example, January 2, 1991, should be written as 01-02-91. Height should be total inches (5' 10" = 70").

The FAA data entry operators processing applications in Oklahoma City are not pilots and they are not acquainted with our idiosyncracies. So the information on the application must be entered in the form they are accustomed to and it must be 100 percent correct or it will be returned and cause a delay or rejection in issuing the permanent pilot certificate with instrument rating.

See FIG. 17-2 for a sample of a correctly completed form. Note that the applicant's height is written as 70 inches.

Finally, be sure your instructor signs his block at the top of the reverse side of the application. If you have forgotten to sign, it's no problem because you can sign in the examiner's office. But if the instructor has forgotten, you may not take the flight test.

THE PRACTICAL TEST

The examiner is responsible for evaluating both the knowledge and the skill of the applicant in meeting the required standards set forth in the practical test standards. Much of the knowledge can be tested during oral questioning, which may take place prior to, or at any time during, the flight test.

It is obviously much more practical to conduct some oral questioning on the ground before the flight test rather than in the air. On the other hand, some questioning is actually more pertinent in the air.

It doesn't make much sense to evaluate an applicant's knowledge of weight and balance computations during the actual flight test. On the other hand, I find I can make much better evaluation of an applicant's knowledge of two-way radio communications failure procedures during the flight by asking, "What would you do now if you suddenly lost two-way radio communications?"

After I have reviewed all your required documents and found them to be in order, I will give you several practical problems to solve. One will be a performance problem, probably computing takeoff roll and obstacle clearance distances on a high density altitude day.

I will ask for a weight and balance calculation. Do yourself a big favor by making blank forms ahead of time for the airplane flown and photocopying two or three for the test. Then all you have to do is fill in the form with the weights and moments of theoretical passengers and baggage the examiner provides and calculate the result. (Be able to do it both ways—the chart method and mathematically.)

You will only have 30 minutes to figure the performance and weight and balance problems *plus* plan an IFR flight to a destination selected by the examiner; this includes planning the flight, preparing a flight log, getting a weather briefing, selecting an alternate, and filling out the flight plan form.

Some intelligent advance planning can save you considerable time (and anxiety) and help you to easily complete all tasks within 30 minutes. Get a full weather briefing for the area within 250 miles before you meet the examiner. Examine instrument approaches to airports within that area that might be suitable for use on the flight test. Study airports and approaches the night before, or even months before, for that matter.

ORAL EXAMINATION

When you present solutions to the performance and weight and balance problems, and the completed flight log and flight plan, I will evaluate your knowledge of these areas.

Other areas of instrument flying in which I will evaluate your knowledge are weather, basic instruments, ATC procedures, FARs, en route and approach charts, and the like. If you have mastered the material covered by Background Briefing 19-20, you should have no trouble with the oral portion of the practical test.

I will also evaluate areas in which you missed written test questions to ensure that you understand them. Even if you scored 100 on the written test, you are going to get some questions on material on which the test is based.

THE FLIGHT TEST

After all this, I'm sure it will be a positive relief to get into the air!

The IFR flight you plan, file, and fly will be to a nearby airport with one of the instrument approaches required for the flight test. During this phase of the flight test I will evaluate your skill in copying clearances and working with ATC during IFR departure, en route, and approach phases. The first instrument approach might be carried through to a landing from a straight-in or a circling approach. Or I might ask you to execute a missed approach, enter the missed approach holding pattern, and then obtain an IFR clearance in the air to a second airport where we will make a second type of instrument approach.

At some point we will cancel IFR (weather permitting) and do some steep turns and unusual attitudes. Count on flying some portion of the flight on partial panel, including an approach.

Expect other simulated emergencies such as electrical failure (simulated by turning off the alternator half of the split power switch) or pitot tube icing (simulated by covering the airspeed indicator).

On occasion I have conducted flight tests in actual IFR flight conditions throughout the entire flight. When the flight plan is filed I will talk to ATC—hopefully with the controller responsible for the sector—and arrange a block of airspace in which to conduct the steep turns and unusual attitudes. What I try to get is a block of air space at an assigned altitude, plus or minus 1,000 feet, along a 20-mile length of airway.

It gets to be a little complicated for the examiner, but the applicant should not assume we won't make the flight if the weather is IFR, or that we won't do steep turns, unusual attitudes, minimum controllable airspeed, stalls, and partial panel under actual IFR conditions.

COMMON DEFICIENCIES

As I covered the other topics in this book I have pointed out several areas in which deficiencies are common on the instrument flight test.

I would say that 90 percent of the applicants I see have their paperwork in order when they show up, previously fewer applicants were properly prepared. Currently, few applicants have difficulty completing their flight planning and the other assigned problems within the 30 minutes allowed. I think the good performance on these items reflects the good work of the flight instructors who, themselves, are apt to be better trained now than in the past.

Holding altitude consistently within the tolerances of the Practical Standards remains a problem and I see the occasional applicant who goes below the minimums on an instrument approach and does not take prompt corrective action.

That is why I always start my instrument students working toward the goal of 2 degrees of heading, 2 knots of airspeed, and 20 feet in altitude from the beginning. If you fly within these tolerances on an instrument flight test, it will be a big boost to your morale and also demonstrate to the examiner that you are a skillful instrument pilot.

NDB procedures continue to be a problem, and it is not uncommon for an applicant to fly an otherwise acceptable instrument test flight, then fail on an NDB approach. This is definitely an area where flight instructors should devote more time and attention.

Another area that requires more attention, I think, is approaches to uncontrolled airports. Sometimes pilots will get so wrapped up in their approaches to an uncontrolled airport that they forget they are about to enter a VFR traffic pattern. They fail to monitor the CTAF, and consequently they are not sure which runway is in use, the local altimeter setting, if available, and if there is other traffic. They fail to notify local traffic that they are arriving on the instrument approach.

It's very uncomfortable for an examiner to sit in the right seat while the applicant goes busting into a VFR traffic pattern with a hood on! The examiner can take control when the safety of the flight is at stake, and this may well be one of those situations. But most of the time the examiner will just have to sit there and let the applicant make mistakes until it becomes pointless to continue.

It's very important for the applicant to realize that he is pilot-in-command on the flight test. The examiner is just an observer. Yes, the examiner is expected to act as a safety pilot while the applicant is under the hood and to take appropriate action if other aircraft, obstructions, or low altitudes are a threat to the safety of the flight, but the applicant is expected to demonstrate his ability as pilot-in-command. The applicant must make the decision whether or not to continue an approach to a landing or to execute a missed approach, for example. The examiner can only interject himself for safety of flight and to carry out the requirements of the practical test standards.

MOVING ON

Earning an instrument rating is an achievement to be proud of and I'm happy to report that most applicants are successful on their first flight test. Serious pilots recognize the high degree of skill and knowledge required to earn this rating, and they respond accordingly. That's certainly one of the reasons I have concentrated on instrument instruction over the years. It's very gratifying to work with highly motivated pilots reaching for another level of excellence.

But the learning process should not end with achievement of the rating. There is always more to learn, and good pilots take advantage of every opportunity to do so.

CURRENCY

To maintain instrument currency you must log at least six hours of instrument time in actual or simulated conditions within the past six calendar months, including at least six instrument approaches. Think of it as 6, 6, and 6. Three of the six hours of instrument flight time must be in the category of aircraft involved. The other three hours, and all six approaches may be in a simulator. See details in FAR 61.57(e).

If these minimum requirements lapse, you must take an instrument competency check (ICC) in order to act as pilot-in-command under IFR. Look upon the ICC as an opportunity to demonstrate and polish skills in areas such as holding patterns, NDB approaches, or DME arcs that you don't have much opportunity to use in the real world of IFR.

The FAA provides no specific guidelines for the ICC, other than that it be flown "in the category of aircraft involved." Theoretically, at least, you could pass an ICC in a Cessna 152, then depart immediately on an actual IFR flight in a Beech Baron. But that wouldn't be too smart and it would defeat the purpose of this important check on your ability to make a safe instrument flight.

When I give an ICC, I treat it like a miniature instrument flight test. I like to make sure the pilot can plan and file an IFR flight, then go out and fly that flight plan in the airplane in which he usually makes IFR flights, working efficiently with ATC. I always try to have the pilot execute the three different approaches required on the flight test—VOR, NDB, and ILS.

Above all, I want to make sure that the pilot is competent and comfortable as pilot-in-command, and that he can confidently handle all the decisions required during an IFR trip. Many sharp pilots take ICCs every six months whether they need them or not in order to maintain a high degree of proficiency.

The biennial flight review (BFR) is another good opportunity to hone your instrument flying skills. You can do an ICC at the same time. This will extend your currency for another six months as well as provide an opportunity to practice instrument procedures that you haven't used recently. The decision is up to the instructor—are you up to the standard?

MOVING UP

The basic principles of instrument flight don't change when you move up to a "high performance airplane," or get a multiengine rating. You still need to scan instruments properly, maintain 2, 2, and 20, and stay ahead of the airplane and the procedures by anticipating the events of the flight while planning and flying.

When you move up to a faster airplane, everything will happen much faster and you must be prepared for it. The best way to prepare is get some instrument instruction in that new high-performance single or light twin from an instructor who is familiar with the airplane.

One of the most important things you can do when moving up is determine the power settings for the new airplane for the different conditions of flight, then automatically set these power settings whenever a change in flight regime is performed.

Determine the best rpm and manifold pressure settings for fast, normal, and slow cruise; normal climb at 500 fpm; descents at 500 and 1,000 fpm; approach, straight-and-level with one increment of flaps down; and approach descent, with one increment of flaps and landing gear down. An actual IFR flight is not the time to discover new knowledge or to experiment.

Write down these settings and always refer to them when changing from one flight condition to another. You can set up the new flight regime very quickly with standard power settings every time. You will avoid throttle jockeying and other forms of overcontrolling. Go right to these standard power settings whenever you make a transition; you can always "fine tune" them or make minor adjustments if you don't get the exact airspeed or rate of climb or descent that you want.

With a multi-engine airplane, take advantage of every opportunity to practice single-engine instrument approaches. Check the operating manual for a "zero thrust" power setting to simulate the drag of a feathered engine.

If you can't find the zero thrust setting in the operating manual, a multi-engine instructor can help you determine one with a little experimentation. With a zero thrust setting you can always have the power of the windmilling engine available during an approach in case of a go-around. If the engine was feathered, a go-around might be impossible.

Consider the single-engine scenario at altitude on an IFR flight plan. If you lose an engine in actual IFR, inform ATC immediately and plan to divert to the closest available airport where the weather is VFR or comfortably above IFR minimums.

You may have to descend to a lower altitude right away. You may have no choice because the airplane will drift down to its single-engine service ceiling no matter what you do. Check the operating manual; the single-engine ceiling may be lower than you think it is.

Light twins frequently come with features you might not be familiar with such as three-axis autopilots, flight directors, and HSIs (horizontal situation indicators). Get checked out by an instructor who is proficient with equipment that is new to you, then become proficient with it yourself by using it whenever possible.

GOOD OPTIONS FOR IFR

As you think ahead to a high-performance single or a light twin, there is some optional equipment you should investigate that will make IFR flying much easier. This equipment is also available for most single-engine airplanes. If you are buying or renting, acquire an airplane that has the optional features you want.

1. **Automatic Pilot with an Approach Coupler.** An autopilot is invaluable on long IFR flights, or when fatigue is a factor. With an autopilot you can give yourself a break. The autopilot can do a major portion of the work if you are re-routed and have to prepare new flight log entries, or if you simply want to eat a sandwich. Because fatigue is worse at the end of an instrument flight, the approach coupler will help relieve the anxiety that often accompanies fatigue and reduce the workload.

2. **DME (distance measuring equipment).** DME is a great aid in visualizing position at all times. Close to a VOR station DME helps you hold a steady reference course until station passage instead of chasing the needle, as you might be tempted to do if you didn't know how close you were. DME holding patterns are easier—you don't have to add or subtract time on the outbound leg to account for the wind—and DME provides access to many approaches that would otherwise be unavailable.

3. **Stormscope.** ATC radar is not designed to detect thunderstorms. If you fly in an area that has frequent thunderstorms, Stormscope can help you work with ATC to deviate around them. This is especially important when coping with embedded thunderstorms that hide along fronts. You can't see them, ATC can't see them, but Stormscope can.

    Airborne weather radar is also effective in spotting thunderstorms. But it is more costly than Stormscope and is impractical for some airplanes.

4. **RNAV (area navigation).** RNAV has matured into an excellent primary navigation system. The most common form of RNAV requires a VORTAC (VOR and DME) receiver, which is less dependent on the vagaries of radio propagation than Loran C, which is not acceptable for IFR unless the equipment and installation is IFR approved. ATC now accepts ''RNAV direct'' flight plans much more willingly than when the equipment was first introduced, and there are a growing number of RNAV approaches throughout the country (FIG. 17-3). The RNAV RWY 32 approach at Greater Buffalo International, New York, is a typical RNAV approach; this is the only authorized straight-in approach to Runway 32 at Buffalo.

5. **DVOR (Digital VOR).** This handy, inexpensive instrument displays a continuous readout of the radial from the VOR or the bearing to it. DVOR takes all the guesswork out of flying direct to a VOR, just turn until your heading matches the magnetic course to the VOR as shown on the DVOR. VOR intersections may be located without changing the OBS settings on navigation receivers.

Just select the radial from the VOR providing the cross-bearing for the fix and watch the numbers change on the DVOR until arrival. If you fly a lot of DME arcs, the DVOR is invaluable, again because it tracks changing radials without resetting the OBS all the time. Most DVORs automatically decode and display the Morse code identifiers.

6. **Personal Computer Briefing Services.** Personal computers are not yet items of airborne equipment, but they offer excellent capabilities in the area of IFR flight planning and I want to mention them.

With a personal computer, a modem, and a subscription to a briefing service provided by such companies as CompuServe and TABS (Total Aviation Briefing Service) you can get printouts of all the information provided by the National Weather Service. The basic service does not require graphics capability on your computer.

Graphics and color capability permit radar summaries and surface analyses with detail and color that surpass TV weather programs. In addition, computer briefing services offer flight planning programs and they can file your flight plans for you.

All these options are wonderful, but you must always remember to aviate, navigate, and communicate without autopilot, RNAV, Stormscope, personal computer, or other advanced equipment.

You must be able to fly hands on, even at the end of a long, fatiguing flight with a lot of weather and turbulence all the way. Many aids can make the flight simpler, easier, more precise, and with less effort. But they are not substitutes for mastering instrument flying.

Only *you* can do that.

Fig. 17-3. *A typical RNAV approach, RNAV RWY 32 at Buffalo, New York.*

246

18
Instrument
Rating Syllabus

FLIGHT LESSON 1: Introduction to IFR

With instructor assistance the student will plan, file, and fly a short instrument cross-country flight to a destination 51 to 75 nautical miles distant and return. The student will learn how to plan an IFR flight, prepare a flight log, obtain a thorough weather briefing, file the IFR flight plans, obtain clearances, and conduct the flight.

The instructor will demonstrate an unhooded instrument approach to a full stop landing at the destination. A new clearance will be obtained for the return flight.

Ideally, the student will be able to fly 60 percent to 90 percent of the flight with some coaching from the instructor.

Assigned Reading:
Chapter 3 - Preparing for an Instrument Flight.
Chapter 4 - Weather/Whether to Fly?
Chapter 5 - How to Get a Good Weather Briefing.
Chapter 6 - Airplane, Equipment, and Instrument Checks.
Chapter 7 - Clearances and Communications.

Preflight Briefing
Review:
1. Communications procedures and frequencies.
Introduce:
1. IFR flight planning.
2. IFR weather briefings.

3. Filing IFR flight plans.
4. Copying clearances.
5. Instrument preflight checks.
6. Departure, en route, and approach procedures.

Completion Standards:

The lesson is complete when the student has a sufficient overview of the planning, filing, and conduct of an IFR flight to begin planning and filing IFR flight plans with a minimum of assistance from the instructor. The student will meet private pilot standards for holding headings within ± 10°, airspeeds within ± 10 knots, and altitudes within ± 100 feet.

Postflight Critique.

BACKGROUND BRIEFING 1-2: Introduction to Basic Instruments

After Flight Lesson 1, the instructor should plan to give several hours of ground instruction covering basic instruments, their purpose, interpretation, and appropriate pilot actions. The emphasis in this briefing is on the relationships and interactions of the control, primary and support instruments and how to scan them properly using the heading, altitude, and attitude scanning pattern.

Use FIG. 8-5, Basic Instruments, and FIG. 12-3, Partial Panel Instruments, as guides to control, primary, and support instruments.

The briefing is complete when the student knows what the instruments show during straight and level flight, turns, climbs, and descents: how control, primary, and support instruments relate to each other during these fundamental maneuvers: and how to scan the instruments properly.

References:

Chapter 6 - Airplane, Instrument, and Equipment Checks.
Chapter 8 - Basic Instruments.
Chapter 9 - Turns, Climbs, and Descents.
Chapter 12 - Stalls, Unusual Attitudes, and Partial Panel.
FAA Instrument Flying Handbook. Chapter II, Instrument Flying: Coping with Illusions in Flight and Chapter IV - Basic Flight Instruments.

Questions:

1. What is the difference between a control instrument and a primary instrument?
2. Give your definition of a primary instrument.
3. Under what conditions would the needle of a primary instrument be moving?
4. What is the primary instrument for:
    Bank during straight and level flight?
    Bank during a standard rate turn?

Bank during a climbing turn?
Bank during a straight ahead descent?
Pitch during level flight?
Pitch during a constant speed climb?
Pitch during a constant rate descent?
Pitch during an ILS approach?
Power during the transition from a climb to straight and level?
Power during a standard rate turn?

5. Describe at least six errors in the use of the heading indicator.
6. Describe at least six common errors in the use of pitch instruments.
7. Describe four common errors in power management.
8. Describe four common errors in the use of trim.
9. The vacuum pump has failed. Describe what happens and what instruments are affected.
10. What causes vertigo? How can you remedy this problem?
11. What causes incorrect airspeed indications? What are preventative measures?
12. Describe the recovery from nose low and nose high unusual attitudes.
13. What indications confirm that you have indeed recovered from an unusual attitude?
14. Describe VOR accuracy requirements prescribed by the FARs and how the regulations may be satisfied.
15. Make a list of the items added to a VFR checklist for the instrument pilot.
16. What are the fuel requirements for an instrument flight?
17. List the equipment checks required by the FARs and when they must be done.
18. What erroneous indications can the attitude indicator show and why?
19. When do you reset the attitude indicator?
20. Describe the errors of the magnetic compass and how to compensate for them.
21. Describe the errors of the VSI and how to compensate.
22. Describe the errors of the fuel gauges and how to compensate.

FLIGHT LESSON 2:
Maneuvering Solely by Reference to Instruments - Part 1

The student will plan, file, obtain the clearance, and depart IFR for the same destination as in Flight Lesson 1. Approximately 20 nautical miles into the flight, the instructor will direct the student to cancel IFR and continue VFR for training in maintaining heading and altitude solely by reference to instruments during straight and level flight, climbs, turns, speed transitions, and descents. The student will monitor a practice IFR approach conducted by the instructor and try to fly as much of it as possible on return to the airport.

Assigned Reading:
Chapter 8 - Basic Instruments.
Chapter 9 - Turns, Climbs, Descents.
Review Chapters 3 through 7.

Preflight Briefing
Review:
1. Flight planning, weather briefing, filing flight plans, copying clearances.
2. Preflight instrument checks.
3. IFR communications.
4. Departure and enroute procedures.
Introduce:
1. Precision straight and level flight.
2. Speed transitions.
3. Standard rate turns.
4. Minimum controllable airspeed
Supplementary Exercises:
1. Pattern A
2. Oboe Pattern

Completion Standards:

The lesson is complete when the student can maintain headings within ± 2 degrees, airspeed within ± 2 knots, altitudes within ± 20 feet for periods of 30 seconds to one minute throughout all maneuvers introduced in this lesson. The student should also begin to develop a general understanding of the instrument approach procedure.
Postflight Critique.

FLIGHT LESSON 3:
Maneuvering Solely by Reference to Instruments—Part II

The student will plan, file, obtain a clearance, and depart on a short IFR cross-country flight to a destination other than that selected for the first two flight lessons. When well-established on the en route portion of the IFR cross-country, the instructor will direct the student to cancel IFR. The student will practice precise heading, altitude, and airspeed control in straight and level flight, climbs, descents, turns, and speed transitions. The student will learn precise heading, altitude, and airspeed control in constant airspeed and constant rate climbs and descents. The student will monitor a practice IFR approach conducted by the instructor on return to the airport.
Assigned Reading:
Review Chapters 3 through 9.
Preflight Briefing
Review:
1. Flight planning, weather briefing, filing flight plans, and copying clearances.
2. Preflight instrument checks.
3. IFR communications.
4. Departure and en route procedures.
5. Straight and level, standard rate turns, and speed transitions.
6. Minimum controllable airspeed.

Introduce:
1. Constant airspeed climbs.
2. Constant rate climbs.
3. Constant airspeed descents.
4. Constant rate descents.

Supplementary Exercises:
1. Step Climbs and descents.
2. Vertical S
3. Pattern B
4. Pattern C

Completion Standards:

The lesson is complete when the student can maintain headings within ± 2 degrees, airspeed within ± 2 knots, and altitudes within ± 20 feet throughout all maneuvers in this lesson for periods of one or two minutes.

Postflight Critique

FLIGHT LESSON 4: VOR Tracking and Bracketing

The student will plan, file, obtain a clearance, and depart on an IFR cross-country flight to a destination suggested by the instructor. The student will practice precise control in straight and level flight, climbs, descents, turns and speed transitions as they occur during the IFR cross-country phase of the flight. When directed by the instructor, the student will cancel IFR.

The student will practice flight at minimum controllable airspeed and practice establishing position in relation to a VOR station. The student will learn to intercept, bracket, and track VOR courses and radials with the needle held within the center circle at all times. The student will monitor a practice instrument approach conducted by the instructor on return to the home airport.

Assigned Reading:
Chapter 10 - VOR Procedures.

Preflight Briefing

Review:
1. Constant airspeed climbs and descents.
2. Minimum controllable airspeed.

Introduce:
1. VOR orientation.
2. VOR tracking and bracketing.

Supplementary Exercises:
1. 16-point orientation exercise.
2. VOR time-distance checks.
3. Pattern A around a VOR.
4. Pattern B around a VOR.

Completion Standards

The lesson is complete when the student can (1) maintain the course deviation indicator needle within one half of its full deflection throughout all the maneuvers in this lesson, except for station passage, and (2) maintain desired headings within ± 2°, airspeed within ± 2 knots, and altitudes within ± 100 feet during the VOR work and then ± 20 feet for extended periods throughout the rest of the flight.

Postflight Critique

FLIGHT LESSON 5: VOR Holding Patterns

The student will file and depart on an IFR cross-country flight to a destination suggested by the instructor using VOR stations as outbound fixes so the student may practice VOR tracking and bracketing; when directed by the instructor, the student will cancel IFR. The student will learn to enter a holding pattern at a VOR station or VOR fix using direct, tear-drop, and parallel methods. The student will learn how to correct for the wind to maintain standard and non-standard holding patterns. On return to the home field, the student will monitor a practice instrument approach conducted by the instructor.

Assigned Reading:

Chapter 11 - Holding Patterns.

Review Chapter 10.

Preflight Briefing

Review:

1. VOR tracking and bracketing, inbound and outbound.

Introduce:

1. Holding pattern entry.

2. Holding patterns.

Completion Standards

The lesson is complete when the student can (1) maintain the course deviation indicator needle within the center circle bull's eye on all inbound courses. (2) maintain inbound legs of holding patterns within ± 15 seconds of the desired one-minute length, and (3) maintain headings within ± 2°, airspeed within ± 2 knots, and altitudes within ± 100 feet during the VOR work and ± 20 feet for extended periods during the rest of the flight.

Postflight Critique.

FLIGHT LESSON 6: Unusual Attitudes, Partial Panel

The student will file and depart on an IFR flight to a destination suggested by the instructor. When directed by the instructor, the student will cancel IFR. The student will practice turns to headings with the magnetic compass. The student will learn power-off stalls and steep turns and will learn to recover from unusual attitudes. The student will learn precise control of the aircraft under partial panel conditions with the attitude indicator and directional indicator covered to simulate a vacuum system failure. The student will fly a no-gyro instrument approach on return to the airport. All of these exercises will be conducted under the hood.

Assigned Reading:
    Chapter 12 - Stalls, Unusual Attitudes, and Partial Panel.
Preflight Briefing
Review:
    1. Turning to a heading with magnetic compass.
Introduce:
    1. Power-off stalls.
    2. Steep turns.
    3. Critical attitude recovery.
    4. Partial panel.
Completion Standards:
    The lesson is complete when the student can (1) recover from power-off stalls with a loss of altitude of 50 feet or fewer, (2) maintain altitude within ± 100 feet, airspeed within ± 10 knots, desired angle of bank within ± 5°, and roll out within ± 10° of the specified heading during 45° banked turns in either direction, (3) recover to straight and level flight without the use of the attitude indicator after an unusual attitude, (4) consistently maintain headings within ± 2°, airspeed within ± 2 knots, and altitudes within ± 20 feet while maneuvering on partial panel.
    The student will demonstrate magnetic compass errors, recognize imminent stalls, and use correct control pressures and movements in proper sequence in unusual attitude recovery in order to complete the lesson.
Postflight Critique.

FLIGHT LESSON 7: ADF Orientation, Tracking, and Bracketing

The student will file and depart on an IFR cross-country flight to a destination selected by the instructor along a route in the vicinity of a strong NDB or standard AM broadcast station. The student will practice control by partial panel in straight and level flight, climbs, turns, descents, and speed transitions as they occur during the IFR cross-country phase of the flight.

The instructor will direct the student to cancel IFR in the vicinity of the NDB. The student will practice steep (45° bank) turns under the hood. The student will learn to orient himself around an NDB and to intercept, bracket, and track inbound and outbound bearings. Pattern A may be used to practice interception, tracking and bracketing.

Assigned Reading:
    Chapter 13 - The NDB Unmasked.
Review:
    1. Partial panel.
    2. Steep turns.
Introduce:
    1. ADF orientation.
    2. ADF tracking and bracketing.

3. Pattern A around an NDB.

4. Pattern B around an NDB.

Supplementary Exercise:

1. ADF time-distance checks.

Completion Standards:

The lesson is complete when the student maintains $\pm$ 100 feet while performing steep turns and is able to predetermine magnetic bearing to an NDB within 10 seconds, then turn to the station and intercept a bearing and track within $\pm$ 5° of the course.

Postflight Critique

FLIGHT LESSON 8: ADF Holding

The student will file and depart on an IFR cross-country flight to a destination selected by the instructor along a route in the vicinity of a strong NDB. When directed by the instructor, the student will cancel IFR and practice ADF orientation, tracking, and bracketing. The student will learn to enter a holding pattern at an NDB by direct, tear-drop, and parallel methods. The student will learn how to correct for the wind to maintain holding patterns.

Assigned Reading:

Review Chapter 13.

Preflight Briefing

Review:

1. Partial panel.

2. ADF orientation.

3. ADF bracketing and tracking.

Introduce:

1. ADF holding patterns.

Completion Standards:

The lesson is complete when the student demonstrates proficiency in predetermining bearing to an NDB, entering holding patterns by the correct method, and intercepting, bracketing, and tracking a magnetic bearing to and from the NDB within $\pm$ 5°.

Postflight Critique

BACKGROUND BRIEFING 8-9: Instrument Approach Procedures

Immediately after completing Background Briefing 1-2, the student should commence work on Background Briefing 8-9, writing as many answers as possible. The student should allow adequate time to prepare for this briefing because it covers a large body of information of great importance, especially on the instrument flight test.

The briefing covers weather minimums, approach charts, alternate airports, communications and clearances, ADF, VOR, and ILS procedures, holding patterns, straight-in and circling approaches, at uncontrolled airports, and missed approaches.

The briefing is complete when the student can talk through ADF, VOR, and ILS approaches—and the appropriate missed approach procedures—at nearby airports.

References:

Chapter 14 - Approaches I: NDB, VOR, and DME.

Chapter 15 - Approaches II: ILS, Localizer, and Radar.

Legends for instrument approach procedure charts.

Questions:

1. Instrument approach procedures are based on criteria established in what publication of the U.S. government?
2. What publication serves as the instrument pilot's reference for transitioning from the en route phase to the landing phase in instrument conditions?
3. Basic weather minimums are prescribed for what two broad categories of approaches?
4. Name six factors that change the published minimums.
5. What does the phrase "vectors to the final approach course" mean?
6. What is the lowest forecast ceiling permitted for an instrument approach at your home airport?
7. How is the highest obstruction at an airport depicted on an approach chart?
8. What is the determining factor in whether or not a legal approach may be attempted?
9. Takeoff weather minimums are found in what publications?
10. Alternate airport weather minimums are found in what publications?
11. When is an alternate airport required? When is an airport authorized as an alternate?
12. In case of communications failure, when and where can an instrument approach be commenced?
13. What are the obstacle clearance altitudes on:
   VOR approaches?
   Localizer approaches?
   ASR approaches?
   NDB approaches?
   DF approaches?
   VOR approaches with FAF?
14. What is the maximum permissible distance from the airport during a circling approach in a Cessna 172?
15. Describe the procedure for executing an early missed approach.
16. Describe holding pattern protection and variations in holding patterns that a pilot might encounter; describe the reasons for these variations.
17. Describe five different acceptable procedures when executing a circling approach.

18. Position reports are not required when in radar contact, except in five specific instances. What are they?
19. When can you descend below the glide slope on an ILS approach?
20. When can you descend below the MDA on a nonprecision approach?
21. Can a takeoff be legally and safely executed when weather is WOXOF?
22. What are you giving up during a WOXOF takeoff ?
23. What are your personal weather minimums for takeoff and for landing? Why?
24. Describe the clearance delivery procedures at your home airport.
25. Describe at least six variations in procedures in receiving IFR clearances.
26. When issued a cruise clearance, how do you get clearance to commence the approach?
27. When in VFR conditions executing an approach, why would you want to cancel IFR?
28. Describe the method of activating pilot-controlled lighting.
29. Describe the purposes of holding patterns depicted by a heavy solid line, a light solid line, and a light dotted line.
30. When stabilized on an ILS approach, you find you are flying 15 knots too fast. What is one popular method of correcting this excess speed?
31. What is the recommended procedure when intercepting the ILS glide slope in a fixed gear, single-engine airplane? In a high performance retractable?
32. When landing at an airport without a control tower, when should you attempt contact on the CTAF? Where do you get the local altimeter setting, and what does it buy you?
33. Where and when should you use that old memory jogger checklist, time, turn, twist, throttle, talk? Explain its use and significance.
34. Must you time an ILS approach? Why? Why not?
35. What special actions must you take when flying into a TCA, restricted area, MOA, and ARSA?
36. Where and how can you tell quickly if your destination airport has an approved instrument approach procedure?
37. Talk your way through several NDB, VOR, ILS, RNAV, LOC/BC, and VOR/DME arc approaches, from the feeder fix through the missed approach.

FLIGHT LESSON 9: ADF Approaches-I

The student will file and fly a short IFR cross-country flight to a nearby destination with a published NDB approach. The student will fly the NDB approach, conduct a missed approach, cancel IFR, then make additional approaches as directed by the instructor. ADF holding patterns in both the approach and the missed approach procedure will be included, if possible.

Assigned Reading:

Chapter 14 - Approaches I: NDB, VOR, and DME.

Review Chapter 13.

Preflight Briefing

Review:

    1. Partial Panel.

Introduce:

    1. ADF approaches.

    2. Missed approaches.

Completion Standards:

The lesson is complete when the student demonstrates an understanding of the NDB approach, tracks the inbound and outbound bearings within ±5°, and maintains altitudes within ±100 feet to MDA. The MDA must be maintained to +100/−0 feet.

Postflight Critique

FLIGHT LESSON 10: ADF Approaches - II

The student will file and fly a short IFR cross-country to a different destination with a published NDB approach. The student will fly the NDB approach, conduct a missed approach, cancel IFR, then make additional NDB approaches, with ADF holding patterns, as directed by the instructor. At least one of the additional NDB approaches will be made with partial panel.

Assigned Reading:

    Review Chapters 13 and 14.

Preflight Briefing

Introduce:

    1. No-gyro approaches.

Review:

    1. Partial panel.

    2. ADF approaches.

    3. Missed approaches.

    4. ADF holding patterns.

Completion Standards:

The lesson is complete when the student demonstrates an understanding of the NDB approach, tracks the inbound and outbound bearings within ±5°, and maintains altitudes within ±100 feet to MDA. The MDA must be maintained to +100/−0 feet.

Postflight Critique

FLIGHT LESSON 11: VOR Approaches - 1

The student will file and fly a short IFR cross-country flight to the nearest destination with a published VOR approach. The student will fly the VOR approach, conduct a missed approach, cancel IFR, then make additional VOR approaches as directed by the instructor.

VOR holding patterns in either the approach or the missed approach procedure will be included.

Assigned Reading:

Review Chapter 14.

Preflight Briefing

Review:

1. Partial panel.

2. Missed approaches.

3. VOR holding patterns.

Introduce:

1. VOR approaches.

Completion Standards:

The lesson is complete when the student demonstrates an understanding of the VOR approach, tracks VOR radials within ±5°, maintains altitudes within ±100 feet to MDA, then +100/−0 feet, and maintains the desired airspeed within ±10 knots.

Postflight Critique

FLIGHT LESSON 12: VOR Approaches - II

The student will file and fly a short IFR cross-country to a different destination with a published VOR approach. The student will fly the VOR approach, conduct a missed approach, cancel IFR, then make additional VOR approaches, with VOR holding patterns, as directed by the instructor. At least one of the additional VOR approaches will be made on partial panel. The student will practice recovery from unusual attitudes.

Assigned Reading:

Review Background Briefing 8-9.

Preflight Briefing

Review:

1. Partial panel.

2. VOR approaches.

3. Missed approaches.

4. VOR holding patterns.

5. Critical attitude recovery.

Completion Standards:

The lesson is complete when the student demonstrates competence in performing VOR approaches and generally tracks VOR radials within ±2°, maintains altitudes within ± 50 feet to MDA, then +50/−0 feet, promptly executes the missed approach, and properly enters a missed approach holding pattern.

Postflight Critique

FLIGHT LESSON 13: VOR Approaches - III

This repeats Flight Lesson 12. The student will file and fly a short IFR cross-country to a different destination with a published VOR approach. The student will fly the VOR approach, conduct a missed approach, cancel IFR, then make additional VOR approaches, with VOR holding patterns, as directed by the instructor. At least one of the additional VOR approaches will be made on partial panel. The student will practice recovery from unusual attitudes.

Assigned Reading:

Review chapters and briefings as directed by instructor.

Preflight Briefing

Review:

    1. Partial panel.

    2. VOR approaches.

    3. Missed approaches.

    4. VOR holding patterns.

    5. Critical attitude recovery.

Supplementary Exercise:

    1. DME arc approaches.

Completion Standards:

The lesson is complete when the student demonstrates competence under full and partial panel conditions and generally tracks VOR radials within $\pm 2°$, maintains altitudes within ± 50 feet to MDA, then $+50/-0$ feet, maintains airspeed within ± 5 knots, promptly executes the missed approach, and properly enters a missed approach holding pattern.

Postflight Critique

FLIGHT LESSON 14: ILS Approaches - I

The student will file and depart on an IFR cross-country flight to a nearby destination with an ILS approach. The student will fly the ILS approach, conduct a missed approach, cancel IFR, then make additional ILS approaches as directed by the instructor. Holding patterns in either the approach or the missed approach procedure will be included.

Assigned Reading:

Chapter 15 - Approaches II: ILS, Localizer, and Radar.

Preflight Briefing

Review:

    1. Missed approaches

Introduce:

    1. ILS approaches

Completion Standards:

The lesson is complete when the student demonstrates an understanding of the ILS approach and maintains altitudes within ± 100 feet, tracks the localizer and glide slope without exceeding full scale deflections, maintains the desired airspeed within ± 10 knots,

and takes prompt action at DH.
Postflight Critique

FLIGHT LESSON 15: ILS Approaches - II

The student will file and fly a short IFR cross-country to a different destination with published ILS and ADF approaches. The student will fly the ILS approach, conduct a missed approach, cancel IFR, then make additional ILS approaches, with holding patterns, as directed by the instructor. At least one of the additional approaches will be made on partial panel.

One of the ILS approaches will be conducted if possible, without the glide slope or using only localizer minimums to simulate the loss of the glide slope receiver. The student will also make one ADF approach to simulate total failure of the aircraft's ILS receiver.

Assigned Reading:
Review Chapter 15.
Preflight Briefing
Review:
1. Partial panel.
2. ILS approaches.
3. ADF approaches.
4. Missed approaches.
Introduce:
1. Loss of radio navigation equipment.
2. Localizer approaches.
Supplementary Exercise:
1. Instrument takeoffs
Completion Standards:
The lesson is complete when the student maintains altitudes within ± 50 feet, tracks localizer and glide slope within ½ scale deflections, maintains desired airspeed within ± 5 knots, recognizes and copes with instrument failures such as loss of glide slope, ILS/localizer, and attitude indicator.
Postflight Critique

FLIGHT LESSON 16:
ILS Back Course, Localizer, LDF, SDF, and Radar Approaches

The student will file and depart on an IFR cross-country to a destination with a published ILS back course approach. The student will fly the ILS back course approach, conduct a missed approach, cancel IFR, then make additional back course approaches, with holding patterns, as directed by the instructor. At least one approach will be made on partial panel.

If an airport with a radar approach is available within the local flying area, a second IFR cross-country should be filed and flown, and several practice radar approaches should be made. If radar approaches are not available, the instructor should simulate the radar approaches by providing vectors and other standard radar approach instructions to a nearby airport.

If airports with SDF and LDA approaches are available within the local flying area, additional IFR cross-country flights should be filed and flown to these airports, time permitting, for practice with these distinctive approaches. Otherwise, they must be covered in ground instruction.

Assigned Reading:

Review chapters and briefings as directed by instructor.

Preflight Briefing

Review:

1. Partial panel.
2. Loss of radio navigation equipment.
3. ILS approaches.
4. Missed approaches.

Introduce:

1. ILS back course approaches.
2. Localizer, LDA, and SDF approaches.
3. Radar Approaches.

Completion Standards:

The lesson is complete when the student demonstrates competence in ILS back course approaches and tracks the localizer within ½ scale deflections, and competence in following directions on radar and no-gyro approaches, maintaining headings within ± 2° and altitudes within ± 50 feet.

Postflight Critique

BACKGROUND BRIEFING 16-17: IFR Cross-Country Procedures

Upon completion of Background Briefing 8-9 the student should commence working on Background Briefing 16-17, again writing as many of the answers as possible. Considerable emphasis is placed upon planning an IFR cross-country flight and its ramifications.

As a designated FAA flight test examiner, I see cross-country planning surfacing again and again as one of the weak areas. Some candidates even expect to conduct their instrument flight test with only a low altitude en route chart, an approach chart or two, and little or no orderliness. A well-organized flight log, thoroughly worked out, is the key to mastering instrument flying.

This briefing is based on the planning for a 200 nautical mile (nm) IFR cross-country. To obtain maximum benefit from the training, it should be a different flight from the 250 nm IFR cross-country flight required by FAR 61.65 (a)(4), which is conducted in Flight Lesson 17.

The briefing is complete when the student is ready in all respects to plan and conduct the 250 nm IFR cross-country required by FAR 61.65 (a)(4).

References:

Chapter 3 - Preparing for an Instrument Flight.

Chapter 17 - IFR Cross-Country Flying.

AIM and A/FD - relevant sections.

En route low altitude chart legend.

Questions:

1. What are four sources of information to determine the IFR route for your flight plan?

2. Plan a 200 nm IFR cross-country flight and work out all the details on your flight log. (Each section of this flight log should be thoroughly explored with your instructor to ensure that you have a complete understanding of the use and value of the flight log.)

3. You have just picked up your clearance and you find that your routing for a segment of the flight plan has been substantially changed. You have at least five courses of action. What are they? When and why would each option be appropriate?

4. Explain the purpose of the TEC and when you would use it.

5. Explain the "Preferred Route" system and how it operates.

6. Give the two main reasons to note the time when you start a takeoff roll on an IFR flight.

7. Give at least two methods for maintaining a record of changes in assigned altitude. Why is this necessary?

8. How are you guaranteed terrain clearance when flying direct (VOR to VOR or off airways)? Who is ultimately responsible?

9. What do the folowing give you? Explain. MEA, MOCA, MCA, MRA, MAA, and MSA

10. Review all the symbols on the en route low altitude chart legend. Ask your instructor to explain any that you don't understand.

11. What is the advantage in recording the time when you reach each fix during an IFR cross-country?

12. What advantage is there in "visualizing" where you are at all times?

13. What is the purpose in writing your clearance limit?

14. Explain when, how, and why you might use the frequency 122.0MHz.

15. How might you use the frequency 121.5MHz? Why? When?

FLIGHT LESSON 17: Long IFR Cross-Country Flight

This flight will satisfy the requirements of FAR 61.65(c)(4) for one cross-country flight "in simulated or actual IFR conditions, on federal airways or as routed by ATC, including one such trip of at least 250 nautical miles, including VOR, ADF and ILS approaches to published minimums at different airports."

The student will file and fly IFR flight plans for each leg of the 250-mile cross-country. Each approach will be made to a full-stop landing, and the student will refile an IFR flight plan after each full-stop landing en route. One approach and landing will be made at an uncontrolled airport.

Assigned Reading:

Chapter 16 - Putting It All Together: The Long IFR Cross-Country.

Preflight Briefing

Review:

1. IFR departure and en route procedures.
2. ADF approaches.
3. VOR approaches.
4. ILS approaches.

Introduce:

1. Lost communications procedures.
2. IFR departures from uncontrolled airports.

Completion Standards:

The lesson is complete when the student demonstrates competence in "putting it all together" and generally maintains airspeed within $\pm$ 2 knots, headings and VOR radials within $\pm$ 2°, and altitudes within $\pm$ 20 feet. The approaches shall meet the FAA practical test standards as stated in the completion standards for Flight Lessons 10, 13, and 15.

Deficiencies on this flight will not normally require repetition of this lesson, but will be corrected with extra work in Flight Lessons 18 and 19.

Postflight Critique

FLIGHT LESSON 18: Progress Check

This flight lesson will be conducted by another instrument flight instructor. The student will file and depart on an IFR cross-country flight to a destination selected by the instructor. When directed by the instructor, the student will cancel IFR and practice maneuvers chosen by the instructor to determine the student's proficiency in carrying out the tasks required by *Instrument Rating Practical Test Standards*. At least one of the instrument approaches will be on partial panel.

Assigned Reading:

As directed by the instructor.

Preflight Briefing

Review:

1. Flight planning.
2. Obtaining and analyzing weather information.
3. Filing an IFR flight plan.
4. IFR departure and en route procedures.
5. Partial panel.
6. Lost communications procedures.

7. Loss of radio navigation equipment.
8. ADF, VOR, and ILS approaches.
9. Holding patterns.
10. Missed approaches.

Completion Standards:

The lesson is complete when the instructor determines what deficiencies, if any, require additional practice.

Postflight Critique

FLIGHT LESSON 19: Flight Test Preparation

The student will file and depart on an IFR cross-country flight to a destination selected by the instructor. When directed by the instructor, the student will cancel IFR and practice any maneuvers that may require further attention to attain the proficiency required by *Instrument Pilot Practical Test Standards*. At least one of the approaches will be on partial panel.

Assigned Reading:

As directed by the instructor.

Preflight Briefing

Review:

1. Flight planning.
2. Obtaining and analyzing weather information.
3. Filing IFR flight plans.
4. IFR departure and en route procedures.
5. Partial panel.
6. Lost communications procedures.
7. Loss of radio navigation equipment.
8. ADF, VOR, and ILS approaches.
9. Holding patterns.
10. Missed approaches.

Completion Standards:

The lesson is complete when the student has corrected any deficiencies noted in Flight Lessons 17 and 18.

Postflight Critique

BACKGROUND BRIEFING 19-20:
Preparation for the Instrument Flight Test Oral Exam

This very important final Background Briefing covers material the student can expect on the oral examination that the designated examiner will give prior to the flight test. Consult Appendix B, "Excerpts from Federal Aviation Regulations," for FARs that deal specifically with IFR trainig and IFR flight. AIM Pilot/Controller Glossary items relating to IFR flight may be found in Appendix C.

The briefing is complete when the student can promptly and accurately answer the questions and work out the problems that can be expected on the oral examination prior to the flight test.

References:

Chapter 17 - How I Conduct an Instrument Flight Test.

Questions:

(Appropriate reference at end of each question.)

1. What are the IFR currency requirements? (FAR 61)
2. If instrument currency expires, how do you become current again? (FAR 61)
3. What is an "appropriate" safety pilot? (FAR 61)
4. What is legally considered "instrument flight time?" (FAR 61)
5. Who has direct responsibility and final authority for the operation of an aircraft? (FAR 91)
6. What certificates must a pilot carry in his personal possession for flight? (FAR 61)
7. What documents must be on board the aircraft for an IFR flight? (FAR 91).
8. What is the fuel requirement for flight in IFR conditions? For VFR flight? (FAR 91)
9. Under what conditions must you list an alternate when filing IFR? (FAR 91)
10. What are alternate airport weather minimums? (FAR 91)
11. What restrictions apply regarding the operation of portable electronic devices on board an aircraft on an IFR flight? (FAR 91)
12. What are the four methods of checking VOR accuracy and the required records? (FAR 91)
13. How often must VOR accuracy be checked for IFR operations? (FAR 91)
14. At what point can you cancel an IFR flight plan? (AIM Chapter 4)
15. How frequently should you check your altimeter setting? (FAR 91)
16. What are the minimum weather conditions for IFR takeoff ? (FAR 91)
17. What additional instruments and equipment are required for IFR over VFR? (FAR 91)
18. Explain DH versus MDA. (FAA *Instrument Flying Handbook*, FAR 1)
19. Explain MEA, MOCA, MRA, MAA. (FAA *Instrument Flying Handbook*, FAR 1)
20. How should you navigate your course on an IFR flight? (FAR 91)
21. Name the components of the ILS system. (AIM Chapter 1)
22. Can anything be substituted for an outer marker on an ILS approach? (AIM Chapter 1)
23. When is a procedure turn prohibited on an instrument approach? (FAR 1, Instrument Approach Procedures legend)
24. When may you descend below DH or MDA? (FAR 91)
25. Explain the terms "straight in" versus "circling" minimums. (FAR 1)
26. How do you determine the minimum safe altitude on a "direct" off-airway flight? (FAR 91)

27. When must the pitot-static system and altimeter be inspected for IFR operations? (FAR 91)

28. How often must the transponder be inspected? (FAR 91)

29. Give the appropriate cruising altitudes when operating IFR below 18,000 feet. (FAR 91)

30. Describe clearance to "VFR on top." (FAR 91, AIM Chapter 4)

31. Do the FARs require an alternate static source? (FAR 91)

32. How will the alternate static source affect the instruments? (FAA *Instrument Flying Handbook*)

33. When should pitot heat be used? When is it recommended? (FAR 91)

34. Explain HAA and HAT. (FAA *Instrument Flying Handbook*, Instrument approach procedures legend)

35. Explain "maintain" versus "cruise" in an IFR assignment. (AIM Pilot/Controller Glossary)

36. List the inspections required on an aircraft to be operated IFR. (FAR 91)

37. Describe the operations and limitations of the gyroscopic instruments. (FAA *Instrument Flying Handbook*)

38. Discuss the purpose and use of SIDs and STARs. (AIM Chapter 4, Pilot/Controller Glossary)

39. Describe a contact approach. (AIM Chapter 4, Pilot/Controller Glossary)

40. Describe a visual approach. (AIM Chapter 4, Pilot/Controller Glossary)

41. Describe minimums as determined by aircraft approach category. How do you know in which category your aircraft is classified? (Instrument approach procedures legend)

42. In a radar environment, what radio reports are expected from you without being requested by ATC? (AIM Chapter 4)

43. Outline your actions if you lose radio communications with ATC. (FAR 91)

44. What would you do if *all* radio equipment failed?

45. How can you determine where restricted areas are located along your route and what are your actions? (En route low altitude chart legend)

46. How can you identify the boundaries between ATC centers? (En route low altitude chart legend)

47. What does the symbol "x" on a flag on an en route chart indicate to the IFR pilot? (En route low altitude chart legend)

48. Describe mandatory changeover points on an airway. What is their purpose? (AIM Chapter 4, En route low altitude chart legend.)

49. Describe the different altitudes shown on an airway. (En route low altitude chart legend)

50. What does the symbol "r" on a flag on an en route chart indicate to the IFR pilot? (En route low altitude chart legend)

51. Describe the mileage markings on an airway. (En route low altitude chart legend)

52. Describe the correct ways to identify intersections. (En route low altitude chart legend)

53. Explain the purpose of aural signals carried by VORs and NDBs and when they are used.

54. What is the significance of a "T" bar on an airway at an intersection versus the absence of such a "T" bar on an airway at an intersection? (En route low altitude chart legend)

55. When can you use DME to identify an intersection? (En route low altitude chart legend)

56. Is there any significance to a solid triangle in the center of a VORTAC? (En route low altitude chart legend)

57. What is the purpose of "Flight Watch?" What is the frequency? (AIM Chapter 6)

58. How can you contact the nearest flight watch as indicated on the chart? (En route low altitude chart legend)

59. What is the difference between holding patterns depicted in fine lines versus those depicted in dark lines on approach charts? (Instrument approach procedures legend)

60. Describe the intent of the circle on approach charts. (Instrument approach procedure chart legend)

61. What are the different ways you might identify an outer marker? (Instrument approach procedure chart legend)

62. How do you identify the highest obstruction on an approach chart? (Instrument approach procedure chart legend)

63. Describe the holding pattern entry on the missed approach at several nearby NDBs. Show examples of where direct, parallel, and tear drop entries must be used.

64. Describe the same entries prescribed in No. 63 on several nearby VOR approaches.

65. Explain feeder routes and show how they are used on several nearby approahces. (Instrument approach procedures chart)

66. Describe how an NDB is depicted on an approach chart and on a low altitude en route chart. (Instrument approach procedures and en route low altitude chart legends)

67. What is the significance of altitudes associated with feeder routes? Show one on a nearby approach? (Instrument approach procedure chart)

68. What is the significance of altitudes marked on the profiles of nearby VOR, VOR/DME, NDB, and ILS approaches? (Instrument approach procedures chart)

69. In case of communications failure, when and where can you begin an approach? (FAR 91)

70. Describe the use of the magnetic compass and common errors associated with this instrument. (FAA *Instrument Flying Handbook*)

71. On an instrument approach, describe the criteria for making a missed approach. (FAR 91)

FLIGHT LESSON 20: Flight Test Recommendation Flight

This flight is a dress rehearsal for the FAA flight test; your instructor will play the role of the FAA examiner. He will direct the student to plan and fly a short IFR cross-country flight. The student will carry out all the tasks specified in the *Instrument Pilot Practical Flight Test Standards*.

Assigned Reading:

Chapter 17 - How I Conduct an Instrument Flight Test.

Preflight Briefing

Review:

1. Flight planning.
2. Obtaining and analyzing weather information.
3. Filing IFR flight plans.
4. IFR departure and en route procedures.
5. Partial panel.
6. Lost communications procedures.
7. Loss of radio navigation equipment.
8. ADF, VOR, and ILS approaches.
9. Holding patterns.
10. Missed approaches.

Completion Standards

The lesson is complete when the instructor is confident the student will pass the FAA instrument flight test and so indicates by endorsing the student's logbook and signing the "Airman Certificate and/or Rating Application," FAA Form 8710-1 (6-86).

Postflight Critique

FLIGHT INSTRUCTOR ENDORSEMENTS

Here is wording approved by the FAA for the various endorsements flight instructors must make in student logbooks to sign a student off for an instrument flight test:

Ground Instruction:

I certify that I have given Mr./Ms. _____ the ground instruction required under FAR 61.65 (b)(1) through (4) for airplanes.

(signed by instructor)
(CFI No. and exp. date)
(date of endorsement)

Flight Instruction:

I certify that I have given Mr./Ms. _____ the flight instruction required under FAR 61.65 (c)(1) through (5) and find him/her competent to take and pass a practical test on these pilot operations.

       (signed by instructor)
       (CFI. No. and exp. date)
       (date of endorsement)

    Instructors will also review written test questions that were missed, as listed on the written test report, and sign and date the report in the space provided to indicate that this review has been completed.

    Finally, the instructor must sign and date the ''Instructor's Recommendation'' portion of the "Airman Certificate and/or Rating Application" (FAA Form 8710-1 (6-86)) (see FIG. 17-2) after checking to make sure that the student has supplied all the required information and filled his portion of the form correctly.

Appendix A

The Instrument Pilot's Professional Library

It is incumbent upon every pilot to develop a professional library, consistent with his personal needs and level of operation. This is particularly important for the instrument pilot.

Following is a bibliography of publications and sources that range from the essential to the "great to have available." Depending on your needs, you may wish to skip some of the items on this list, preferring to consult these references as they might be needed from some aviation library or from other pilots. But it is good to know what is available.

This set of publications will quickly become something you will turn to again and again to refresh your memory about something you haven't encountered, or to broaden your knowledge on subjects such as weather.

A good professional library will contain all the information on which written, oral, and flight tests are based. You must read these publications and understand what they cover, especially on such subjects as air traffic control procedures, instrument construction and behavior, theory of flight, weather, and weather reports.

The seven basic books and two subscriptions listed immediately below are all available through the U.S. Government Printing Office (AC stands to Advisory Circular):

Airman's Information Manual (AIM), Official Guide to Basic Flight Information and ATC Procedures. This publication is issued every 16 weeks by the FAA and contains basic fundamentals required to fly in the U.S. National Airspace System. Among other data, it also contains adverse factors affecting safety of flight; health and medical facts of interest to pilots; ATC information affecting rules, regulations, and procedures; a pilot/controller glossary; Air Defense Identification Zones (ADIZ); Designated Mountain Areas; and emergency procedures. Annual subscription $20, stock number SN 750-001-00000-9.

Aviation Weather, AC 00-6A - Provides an up-to-date and expanded text for pilots and other flight operations personnel whose interest in meteorology is primarily in its application to flying. $8.50, stock number SN 050-007-00283-1.

FAA Aviation News. A bi-monthly report issued by the FAA giving air traffic control and safety news, current interpretations of FARs, and general information. Annual subscription $5.50, stock number SN 750-002-00000-5.

Aviation Weather Services, AC 00-45C - Supplements AC 00-6A, *Aviation Weather,* in that it explains the weather service in general and the use and interpretation of reports, forecasts, weather maps, and prognostic charts in detail. $6, stock number SN 050-007-00705-1.

Instrument Flying Handbook, AC 61-27C - Provides the pilot with basic information needed to acquire an FAA instrument rating. It is designed for the reader who holds at least a Private Pilot Certificate and is knowledgeable in all areas covered in the *Pilot's Handbook of Aeronautical Knowledge.* $8.50, stock number SN 050-007-00585-7.

Instrument Rating Question Book, FAA-T-8080-7B - Developed by the FAA for use by applicants who are preparing for the Instrument Rating Written Test. It serves as an official test book as well as a test guide for studying and preparing for the Instrument Written Test. $9, stock number SN 050-007-00784-1.

(There are also several excellent commercial reprints of this publication that include answers and explanations. These commercial publications with answers and explanations will be more helpful in preparing for the Instrument Written Exam; I recommend them.)

Instrument Rating Practical Test Standards, FAA-S-8081-4 - Contains the standards which establish the practical test requirements for instrument rating certification in all aircraft categories and associated classes. Used by FAA inspectors and designated pilot examiners when conducting airman practical tests (oral and practical). Flight instructors and applicants will find these standards helpful in practical test preparation. Annual subscription $14, stock number SN 950-060-00000-0. (Commercial reprints are available from FBOs and other sources on a non-subscription basis.)

Medical Handbook for Pilots, AC 67-2 - An aviation medicine handbook written in pilots' language that provides guidance on when, and when not to fly. This book emphasizes the fact that to be a good pilot you must be physically fit, psychologically sound, and well trained. Designed to complement the *Pilot's Handbook of Aeronautical Knowledge.* $5, stock number SN 050-007-00254-8.

Pilot's Handbook of Aeronautical Knowledge, AC 61-23B - Contains essential information used in training and guiding applicants for pilot certification, flight instructors, and flying school staffs. $10, stock number SN 050-011-00077-7.

Order the publications above from:

Superintendent of Documents
U.S. Government Printing Office
Washington, D.C. 20402

You can copy the form in FIG. A-1 and use it for single publications. Copy and use the form in FIG. A-2 for subscriptions. Note that both these forms have provisions for charging your order to a credit card. You can also purchase reprints of many of these publications at FBOs and through many aviation supply outlets.

The FAA's master guide to publications is Advisory Circular AC 00-2, "Advisory Circular Checklist (and Status of Other FAA Publications)," which is updated annually. It gives the current status and ordering instructions for all of the publications above plus many others published by the current update of AC 00-2 will be sent to you free of charge on request. Free Advisory Circulars may be obtained by requesting them through:

U.S. Department of Transportation
Utilization and Storage Section, M443.2
Washington, D.C. 20590

See FIG. A-3 for a convenient order form provided by the FAA that you can copy. Using it will speed up the process. When the ACs come in, put them in a ring binder.

While you're at it, there are some other free ACs that you should order at the same time. Be persistent; if something you want is reported as out of stock or being revised try again at a later date. Here are the free ACs that every instrument pilot should be familiar with:

AC 00-24B - *Thunderstorms.* Describes the hazards of thunderstorms to aviation and offers guidance to help prevent accidents caused by thunderstorms.

AC 00-44Z - *Status of the Federal Aviation Regulations.* This circular sets forth the current publication status of Federal Aviation Regulations, any changes to date, and provides a price list.

AC 20-3C - *Status and Availability of Military Handbooks and ANC Bulletins for Aircraft.* Announces the status and availability of military handbooks and ANC bulletins prepared jointly by the FAA, Navy, and Air Force.

AC 60-11A - *Aids Authorized for Use by Airmen Written Test Applicants.* Clarifies FAA policy concerning aids that applicants may use when taking airman written tests.

Appendix A

Superintendent of Documents Publication Order Form

Order Processing Code

*

Charge your order.
It's easy!   MasterCard   VISA   CHOICE

1. Please Type or Print (Form is aligned for typewriter use.)

All prices include regular domestic postage and handling and are good through   After this date, please call Order and
Information Desk at 202-783-3238 to verify prices.

Qty.	Stock Number	Title	Price Each	Total Price

Please Type or Print   International customers please add an additional 25%   Total for Publications

2. _____
(Company or personal name)

_____
(Additional address/attention line)

_____
(Street address)

_____
(City, State, ZIP Code)

(___) _____
(Daytime phone including area code)

3. Please Choose Method of Payment:

☐ Check payable to the Superintendent of Documents

☐ GPO Deposit Account ⬚⬚⬚⬚⬚⬚⬚ — ⬚

☐ VISA, CHOICE or MasterCard Account

⬚⬚⬚⬚⬚⬚⬚⬚⬚⬚⬚⬚⬚⬚⬚⬚⬚⬚⬚⬚

_____ (Credit card expiration date)   *Thank you for your order!*

_____
(Signature)

4. Mail To: Superintendent of Documents, Government Printing Office, Washington, D.C. 20402-9325

Superintendent of Documents Publication Order Form

Order Processing Code

*

Charge your order.
It's easy!   MasterCard   VISA   CHOICE

1. Please Type or Print (Form is aligned for typewriter use.)

All prices include regular domestic postage and handling and are good through   After this date, please call Order and
Information Desk at 202-783-3238 to verify prices.

Qty.	Stock Number	Title	Price Each	Total Price

Please Type or Print   International customers please add an additional 25%   Total for Publications

2. _____
(Company or personal name)

_____
(Additional address/attention line)

_____
(Street address)

_____
(City, State, ZIP Code)

(___) _____
(Daytime phone including area code)

3. Please Choose Method of Payment:

☐ Check payable to the Superintendent of Documents

☐ GPO Deposit Account ⬚⬚⬚⬚⬚⬚⬚ — ⬚

☐ VISA, CHOICE or MasterCard Account

⬚⬚⬚⬚⬚⬚⬚⬚⬚⬚⬚⬚⬚⬚⬚⬚⬚⬚⬚⬚

_____ (Credit card expiration date)   *Thank you for your order!*

_____
(Signature)

4. Mail To: Superintendent of Documents, Government Printing Office, Washington, D.C. 20402-9325

Fig. A-1. *Form for ordering non-subscription publications from Government Printing Office.*

274

Superintendent of Documents Subscriptions Order Form

Order Processing Code:
∗

Charge your order.
It's easy! MasterCard VISA CHOICE

1. Please Type or Print (Form is aligned for typewriter use.)

All prices include regular domestic postage and handling and are subject to change. International customers please add an additional 25%.

Qty.	List ID	Title	Price Each	Total Price
		Total for Subscriptions		

Please Type or Print

2. _____
(Company or personal name)

_____
(Additional address/attention line)

_____
(Street address)

_____
(City, State, ZIP Code)

( ) _____
(Daytime phone including area code)

3. Please Choose Method of Payment:

☐ Check payable to the Superintendent of Documents

☐ GPO Deposit Account ⬜⬜⬜⬜⬜⬜–⬜

☐ VISA, CHOICE or MasterCard Account

⬜⬜⬜⬜⬜⬜⬜⬜⬜⬜⬜⬜⬜⬜⬜⬜⬜⬜⬜⬜

_____
(Credit card expiration date)   *Thank you for your order!*

_____
(Signature)

4. Mail To: Superintendent of Documents,
Government Printing Office, Washington, D.C. 20402-9371

Superintendent of Documents Subscriptions Order Form

Order Processing Code:
∗

Charge your order.
It's easy! MasterCard VISA CHOICE

1. Please Type or Print (Form is aligned for typewriter use.)

All prices include regular domestic postage and handling and are subject to change. International customers please add an additional 25%.

Qty.	List ID	Title	Price Each	Total Price
		Total for Subscriptions		

Please Type or Print

2. _____
(Company or personal name)

_____
(Additional address/attention line)

_____
(Street address)

_____
(City, State, ZIP Code)

( ) _____
(Daytime phone including area code)

3. Please Choose Method of Payment:

☐ Check payable to the Superintendent of Documents

☐ GPO Deposit Account ⬜⬜⬜⬜⬜⬜–⬜

☐ VISA, CHOICE or MasterCard Account

⬜⬜⬜⬜⬜⬜⬜⬜⬜⬜⬜⬜⬜⬜⬜⬜⬜⬜⬜⬜

_____
(Credit card expiration date)   *Thank you for your order!*

_____
(Signature)

4. Mail To: Superintendent of Documents,
Government Printing Office, Washington, D.C. 20402-9371

Fig. A-2. *Form for ordering subscriptions from Government Printing Office.*

Appendix A

ORDER BLANK (Free Publications)   DATE ____ / ____ / ____

For Faster Service Use A Self-Addressed Mailing
Label when not using this blank. Please Print
Or Type All Information.

Mail TO:

DOT, M-443.2
UTILIZATION AND STORAGE SECTION
WASHINGTON, D.C. 20590

NUMBER	TITLE	QTY.	CODE	____ / *Action Code Months Before Reordering (Codes 1 & 2)

Request Filled By: _____ DATE ____ / ____ / ____

* **Action Codes** (to be entered in action column by supply personnel).

1. Out of Stock **(reorder in 60 days)** * *   4. Canceled by _____ **(enclosed)**

2. Being revised   5. Other: _____

3. Cancelled, **no replacement**

* *IF YOU DO NOT RECEIVED DESIRED PUBLICATION (S) AFTER YOUR <u>SECOND</u> REQUEST
PLEASE CALL FAA'S TOLL-FREE CONSUMER HOTLINE: **1 (800) FAA-SURE.**

TO COMPLETE ORDER: Enter Name and Address.   <u>Do Not Detach.</u>

NAME
_____

STREET ADDRESS
_____

CITY	STATE	ZIP CODE

Fig. A-3. *Form for ordering free FAA Advisory Circulars (ACs) from Department of Transportation.*

AC 90-42D - *Traffic Advisory Practices at Nontower Airports.* This circular contains good operating practices and procedures for use when approaching and departing airports that do not have a tower in operation.

AC 91-46 - *Gyroscopic Instruments—Good Operating Practices.* Issued to re-emphasize to general aviation instrument-rated pilots the need to determine the proper operation of gyroscopic instruments, the importance of instrument cross-checks, and proficiency in partial panel operations.

AC 91.83-1B - *Cancelling or Closing Flight Plans.* Explains FAA Procedure for Handling flight plans and emphasizes the need for cancelling or closing flight plans.

AC 97-1A - *Runway Visual Range.* Describes RVR measuring equipment and its operating use.

The FAA maintains a mailing list of people to whom they routinely send updated or reissued free ACs. You can get on this list by sending your name and address to:

Department of Transportation
Distribution Requirements Section, M-494.1
400 Seventh Street SW
Washington, D.C. 20590

State in your request the AC categories you wish to receive free. The most useful categories of free ACs for instrument pilots are:

00 - General
60 - Airmen
90 - Air Traffic and General Operating Rules

If you do not receive the publications you want, or need assistance in ordering, the FAA has a consumer hotline that will help you. Call 1-800-FAA-SURE (1-800-322-7873).

For obvious reasons, the use of obsolete charts or publications for navigation may be dangerous. Aeronautical information changes rapidly, and it is vitally important that pilots check the effective dates on each aeronautical chart and publication to be used. Obsolete charts should always be discarded and replaced by current editions.

Because you can't always rely on an FBO or flight school to have exactly what you need, you should have your own subscriptions to the *Airport/Facility Directory* (the green book), a set of en route charts that covers the region in which you fly, and your own current booklets of NOS Approach, SID, and STAR charts. For prices, subscription rates, and ordering instructions, get a copy of the latest free catalog, *Aeronautical Charts and*

Appendix A

Related Products, published by the National Oceanic and Atmospheric Administration (NOAA). You can get this catalog by writing to:

NOAA Distribution Branch, N/CG33
National Ocean Service
Riverdale, Maryland 20737

The NOAA Distribution Branch may also be reached at the following telephone numbers:

General information and one-time sales: 301/436-6990
Subscriptions: 301/436-6993
Chart agent orders and information: 301/436-8726

Information on Canadian charts and publications may be obtained from:

Canada Map Office
Department of Energy, Mines, and Resources
615 Booth Street
Ottawa, Ontario, K1A 0E9
613/998-3865

Information on Jeppesen charts may be obtained by contacting:

Jeppesen Sanderson
55 Inverness Drive East
Englewood, Colorado 80112-5498
303/799-9090

A very readable and informative publication of interest to all pilots is the *Flight Instructors' Safety Report* published by the AOPA Air Safety Foundation four times a year, in January, April, July, and October. Flight instructors receive the publication free of charge, others may obtain an annual subscription for $20. If you make a donation to the Air Safety Foundation of $30 or more, the publication will be sent to you on request. Contact:

AOPA Air Safety Foundation
421 Aviation Way
Frederick, Maryland 21701
1-800-638-3101

Instrument flight instructors and instrument pilots who wish to delve deeper into the real world of instrument flying should also obtain these publications:

Aeronautical Chart User's Guide. This is a guide to the wealth of information provided on NOAA's aeronautical charts, both VFR and IFR. It includes a discussion of IFR chart terms and symbols and complete, illustrated lists of all symbols used on all NOAA charts.

278

Price: $4. See listing in "Aeronautical Charts and Related Products" above for ordering instructions.

Air Traffic Control Handbook, 7110.65E. Prescribes air traffic control procedures and phraseology for use by air traffic control personnel. Controllers are required to be familiar with all provisions of this handbook. Annual subscription $40, GPO stock number SN 950-002-00000-0.

Contractions, 7340.1K. Gives approved word and phrase contractions used by personnel connected with air traffic control, communications, weather, charting, and associated services. Annual subscription $28, GPO stock number SN 950-003-00000-6.

Flight Services, 7110.10H. Prescribes procedures and phraseology for use by personnel providing flight assistance services. Annual subscription $46, GPO stock number SN 950-032-00000-6.

Guide to Federal Aviation Administration Publications, FAA-APC-PG-10. A 60-page book, with order blanks, that explains how the whole range of FAA documents and publications can be obtained. Lists publications by category and gives the various sources. Updated annually. Free, send order to Public Inquiry Center, APA-230, Federal Aviation Administration, Washington, D.C. 20591.

United States Standard for Terminal Instrument Procedures (TERPS), 8260.3B. Contains criteria which shall be used to formulate, review, approve, and publish procedures for instrument approach and departure of aircraft to and from civil and military airports. Price: $7, GPO stock number SN 050-007-00345-5.

Appendix B

Excerpts from Federal Aviation Regulations

FARs 1, 61, and 91 are fundamental regulations for general aviation, and every pilot should be familiar with all their sections and subsections. Excerpted here are only those pertinent to instrument flight.

1.1 General Definitions

Airport traffic area means, unless otherwise specifically designated in Part 93, that airspace within a horizontal radius of 5 statute miles from the geographic center of any airport at which a control tower is operating, extending from the surface up to, but not including, an altitude of 3,000 feet above the elevation of the airports.

Air traffic means aircraft operating in the air or on an airport surface, exclusive of loading ramps and parking areas.

Air traffic clearance means an authorization by air traffic control, for the purpose of preventing collision between known aircraft, for an aircraft to proceed under specified traffic conditions within controlled airspace.

Air traffic control means a service operated by appropriate authority to promote the safe, orderly, and expeditious flow of air traffic.

Alternate airport means an airport at which an aircraft may land if a landing at the intended airport becomes inadvisable.

Area navigation (RNAV) means a method of navigation that permits aircraft operations on any desired course within the coverage of station-referenced navigation signals or within the limits of self-contained system capability.

Area navigation high route means an area navigation route within the airspace extending upward from, and including, 18,000 feet MSL to Flight Level 450.

Appendix B

Area navigation low route means an area navigation route within the airspace extending upward from 1,200 feet above the surface of the earth to, but not including 18,000 feet MSL.

Category II operation, with respect to the operation of aircraft, means a straight-in ILS approach to the runway of an airport under a Category II ILS instrument approach procedure issued by the Administrator or other appropriate authority.

Category III operations, with respect to the operation of aircraft, means an ILS approach to, and landing on, the runway of an airport using a Category III ILS instrument approach procedure issued by the Administrator or other appropriate authority.

Ceiling means the height above the earth's surface of the lowest layer of clouds or obscuring phenomena that is reported as broken, overcast, or obscuration, and not classified as thin or partial.

Decision height, with respect to the operation of aircraft, means the height at which a decision must be made, during an ILS or PAR instrument approach, to either continue the approach or to execute a missed approach.

Flight level means a level of constant atmospheric pressure related to a reference datum of 29.92 inches of mercury. Each is stated in three digits that represent hundreds of feet. For example, Flight Level 250 represents a barometric altimeter indication of 25,000 feet: Flight Level 255, an indication of 25,500 feet.

Flight plan means specified information, relating to the intended flight of an aircraft, that is filed orally or in writing with air traffic control.

Flight time means the time from the moment the aircraft first moves under its own power for the purpose of flight until the moment it comes to rest at the next point of landing. (Block-to-block time.)

Flight visibility means the average forward horizontal distance, from the cockpit of an aircraft in flight, at which prominent unlighted objects may be seen and identified by day and prominent lighted objects may be seen and identified by night.

IFR conditions means weather conditions below the minimum for flight under visual flight rules.

IFR over-the-top, with respect to the operation of aircraft, means the operation of an aircraft over-the-top on an IFR flight plan when cleared by air traffic control to maintain VFR conditions or VFR conditions on top.

Minimum descent altitude means the lowest altitude, expressed in feet above mean sea level, to which descent is authorized on final approach or during circle-to-land maneuvering in execution of a standard instrument approach procedure, where no electronic glide slope is provided.

Night means the time between the end of evening civil twilight and the beginning of morning civil twilight, as published in the American Air Almanac, converted to local time.

Nonprecision approach procedure means a standard instrument approach procedure in which no electronic glide slope is provided.

Over-the-top means above the layer of clouds or other obscuring phenomena forming the ceiling.

Pilot in command means the pilot responsible for the operation and safety of an aircraft during flight time.

Positive control means control of all air traffic, within designated airspace, by air traffic control.

Precision approach procedure means a standard instrument approach procedure in which an electronic glide slope is provided.

RNAV way point (W/P) means a predetermined geographical position used for route or instrument approach definition or progress reporting purposes that is defined relative to a VORTAC station position.

Route segment means a part of a route. Each end of that part is identified by: (1) a continental or insular geographical location; (2) a point at which a definite radio fix can be established.

True airspeed means the airspeed of an aircraft relative to undisturbed air.

VFR over-the-top, with respect to the operation of aircraft, means the operation of an aircraft over-the-top under VFR when it is not being operated on an IFR flight plan.

1.2 Abbreviations and Symbols

AGL	above ground level.
ALS	approach light system.
ASR	airport surveillance radar.
ATC	air traffic control.
DH	decision height.
DME	distance measuring equipment compatible with TACAN.
FM	fam marker.
GS	glide slope.
HIRL	high-intensity runway light.
IAS	indicated airspeed.
IFR	instrument flight rules.
ILS	instrument landing system.
IM	ILS inner marker.
INT	intersection.
LDA	localizer-type directional aid.
LMM	compass locator at middle marker.
LOC	ILS localizer.
LOM	compass locator at outer marker.
MAA	maximum authorized IFR altitude.
MALS	medium intensity approach light system.
MALSR	medium intensity approach light system with runway alignment indicator lights.
MCA	minimum crossing altitude.
MDA	minimum descent altitude.
MEA	minimum en route IFR altitude.

MM	ILS middle marker.
MOCA	minimum obstruction clearance altitude.
MRA	minimum reception altitude.
MSL	mean sea level.
NDBADF	nondirectional beacon automatic direction finder.
NOPT	no procedure turn required.
OM	ILS outer marker.
PAR	precision approach radar.
RAIL	runway alignment indicator light system.
RBN	radio beacon.
RCLM	runway centerline marking.
RCLS	runway centerline light system.
REIL	runway end identification lights.
RVR	runway visual range as measured in the touchdown zone area.
SALS	short approach light system.
SSALS	simplified short approach light system.
SSALSR	simplified short approach light system with runway alignment indicator lights.
TACAN	ultra-high frequency tactical air navigational aid.
TAS	true airspeed.
TDZL	touchdown zone lights.
TVOR	very high frequency terminal omnirange station.
VFR	visual flight rules.
VHF	very high frequency.
VOR	very high frequency omnirange station.
VORTAC	collocated VOR and TACAN.

PART 61 - CERTIFICATION: Pilots and Flight Instructors

61.1 Applicability

(a) This Part prescribes the requirements of issuing pilot and flight instructor certificates and ratings, the conditions under which those certificates and ratings are necessary, and the privileges and limitations of those certificates and ratings.

61.3 Requirements for Certificates, Rating, and Authorizations

(e) *Instrument rating.* No person may act as pilot in command of a civil aircraft under instrument flight rules, or in weather conditions less than the minimum prescribed for VFR flight unless:

    (1) In the case of an airplane, he holds an instrument rating or an airline transport pilot certificate with an airplane category rating on it.

61.43 Flight Tests: General Procedures

(a) The ability of an applicant for a private or commercial pilot certificate, or for an aircraft or instrument rating on that certificate to perform the required pilot operations is based on the following:

    (1) Executing procedures and maneuvers within the aircraft's performance capabilities and limitations, including use of the aircraft's systems.

    (2) Executing emergency procedures and maneuvers appropriate to the aircraft.

    (3) Piloting the aircraft with smoothness and accuracy.

    (4) Exercising judgment.

    (5) Applying his aeronautical knowledge.

    (6) Showing that he is the master of the aircraft, with the successful outcome of a procedure or maneuver never seriously in doubt.

61.51 Pilot Logbooks

(c) Logging of pilot time—

    (3) *Second-in-command flight time.* A pilot may log as second in command time all flight time during which he acts as second in command of an aircraft on which more than one pilot is required under the type certification of the aircraft, or the regulations under which the flight is conducted.

    (4) *Instrument flight time.* A pilot may log as instrument flight time only that time during which he operates the aircraft solely by reference to instruments, under actual or simulated instrument flight conditions. Each entry must include the place and type of each instrument approach completed, and the name of the safety pilot for each simulated instrument flight. An instrument flight instructor may log as instrument time that time during which he acts as instrument flight instructor in actual instrument weather conditions.

61.55 Second-in-Command Qualifications

(a) Except as provided [elsewhere] in this section, no person may serve as second in command of an aircraft type certified for more than one required pilot flight crewmember unless that person holds—

    (1) At least a current private pilot certificate with appropriate category and class ratings; and

    (2) An appropriate instrument rating in the case of flight under IFR.

61.57 Recent Flight Experience: Pilot in Command

(e) *Instrument.*

    (1) *Recent IFR experience.* No pilot may act as pilot in command under IFR, nor in weather conditions less than the minimums prescribed for VFR, unless he has, within the past 6 calendar months—

(i) In the case of an aircraft other than a glider, logged at least 6 hours of instrument time under actual or simulated IFR conditions, at least 3 of which were in flight in the category of aircraft involved, including at least 6 instrument approaches, or passed an instrument competency check in the category of aircraft involved.

(2) *Instrument competency check.* A pilot who does not meet the recent instrument experience requirements of subparagraph (1) of this paragraph during the prescribed time or 6 calendar months thereafter may not serve as pilot in command under IFR, nor in weather conditions less than the minimums prescribed for VFR, until he passes an instrument competency check in the category of aircraft involved, given by an FAA inspector, a member of an armed force in the United States authorized to conduct flight tests, an FAA-approved check pilot, or a certificated instrument flight inspector. The Administrator may authorize the conduct of part or all of this check in a pilot ground trainer equipped for instruments or an aircraft simulator.

61.65 Instrument Rating Requirements

(a) *General.* To be eligible for an instrument rating (airplane) or an instrument rating (helicopter), an applicant must:

(1) Hold at least a current private pilot certificate with an aircraft rating appropriate to the instrument rating sought;

(2) Be able to read, speak, and understand the English language; and

(3) Comply with the applicable requirements of this section.

(b) *Ground instruction.* An applicant for the written test for an instrument rating must have received ground instruction, or has logged home study in at least the following areas of aeronautical knowledge appropriate to the rating sought.

(1) The regulations of this chapter that apply to flight under IFR conditions, the *Airman's Information Manual,* and the IFR air traffic system and procedures;

(2) Dead reckoning appropriate to IFR navigation, IFR navigation by radio aids using the VOR, ADF, and ILS systems, and the use of IFR charts and instrument approach plates;

(3) The procurement and use of aviation weather reports and forecasts, and the elements of forecasting weather trends on the basis of that information and personal observation of weather conditions; and

(4) The safe and efficient operation of airplanes or helicopters, as appropriate, under instrument weather conditions.

(c) *Flight instruction and skill—airplanes.* An applicant for the flight test for an instrument rating (airplane) must present a logbook record certified by an authorized flight instructor showing that he has received instrument flight instruction in an airplane in the following pilot operations, and has been found competent in each of them:

    (1) Control and accurate maneuvering of an airplane solely by reference to instruments.

    (2) IFR navigation by the use of the VOR and ADF systems, including the compliance with air traffic control instructions and procedures.

    (3) Instrument approaches to published minimums using the VOR, ADF, and ILS system (instruction in the use of the ADF and ILS may be received in an instrument ground trainer and instruction in the use of the ILS glide slope may be received in an airborne ILS simulator).

    (4) Cross-country flying in simulated or actual IFR conditions, on Federal airways or as routed by ATC, including one such trip of at least 250 nautical miles, including VOR, ADF, and ILS approaches at different airports.

    (5) Simulated emergencies, including the recovery from unusual attitudes, equipment or instrument malfunctions, loss of communications, and engine-out emergencies if a multiengine airplane is used, and missed approach procedure.

  (d) *Instrument instruction and skill—(helicopter)*. An applicant for the flight test for an instrument rating (helicopter) must present a logbook record certified to by an authorized flight instructor showing that he has received instrument flight instruction in a helicopter in the following pilot operations and has been found competent in each of them:

    (1) The control and accurate maneuvering of a helicopter solely by reference to instruments.

    (2) IFR navigation by the use of the VOR and ADF systems, including compliance with air traffic control instructions and procedures.

    (3) Instrument approaches to published minimums using the VOR, ADF, and ILS systems (instruction in the use of the ADF and ILS may be received in an instrument ground trainer, and instruction in the use of the ILS glide slope may be received in an airborne ILS simulator).

    (4) Cross-country flying under simulated or actual IFR conditions, on Federal airways or as routed by ATC, including one flight of at least 100 nautical miles, including VOR, ADF, and ILS approaches at different airports.

    (5) Simulated IFR emergencies, including equipment malfunctions, missed approach procedures, and deviations to unplanned alternates.

  (e) *Flight experience*. An applicant for an instrument rating must have at least the following flight time as a pilot:

    (1) A total of 125 hours of pilot flight time, of which 50 hours are as pilot in command in cross-country flight in a powered aircraft with other than a student pilot certificate. Each cross-country flight must have a landing at a point more than 50 nautical miles from the original departure point.

    (2) 40 hours of simulated or actual instrument time, of which not more than 20 hours may be instrument instruction by an authorized instructor in an instrument ground trainer acceptable to the Administrator.

(3) 15 hours of instrument flight instruction by an authorized flight instructor, including at least 5 hours in an airplane or a helicopter, as appropriate.

(f) *Written test.* An applicant for an instrument rating must pass a written test appropriate to the instrument rating sought on the subjects in which ground instruction is required by paragraph (b), of this section.

(g) *Practical test.* An applicant for an instrument rating must pass a flight test [based upon the *Instrument Rating Practical Test Standards,* FAA-S-8081-4, 1986 plus subsequent revisions.] in an airplane or a helicopter, as appropriate. The test must include instrument flight procedures selected by the inspector or examiner conducting the test to determine the applicant's ability to perform competently the IFR operations on which instruction is required by paragraph (c) or (d) of this section.

PART 91—GENERAL OPERATING AND FLIGHT RULES

91.21 Flight Instruction; Simulated Instrument Flight and Certain Flight Tests

(b) No person may operate a civil aircraft in simulated instrument flight unless:
  (1) An appropriately rated pilot occupies the other control seat as safety pilot;
  (2) The safety pilot has adequate vision forward and to each side of the aircraft, or a competent observer in the aircraft adequately supplements the vision of the safety pilot; and
  (3) Except in the case of lighter-than-air aircraft, that aircraft is equipped with fully functioning dual controls. However, simulated instrument flight may be conducted in a single-engine airplane, equipped with a single, functioning, throw over control wheel, in place of fixed, dual controls of the elevator and ailerons, when:
    (i) The safety pilot has determined that the flight can be conducted safely; and
    (ii) The person manipulating the control has at least a private pilot certificate with appropriate category and class ratings.

91.23 Fuel Requirements for Flight in IFR Conditions

(a) Except as provided in paragraph (b) of this section, no person may operate a civil aircraft in IFR conditions unless it carries enough fuel (considering weather reports and forecasts, and weather conditions) to:
  (1) Complete the flight to the first airport of intended landing;
  (2) Fly after that for 45 minutes at normal cruising speed.
  (3) Fly after that for 45 minutes at normal cruising speed.
(b) Paragraph (a) (2) of this section does not apply if:

(1) Part 97 of this subchapter prescribes a standard instrument approach procedure for the first airport of intended landing; and

(2) For at least 1 hour before and 1 hour after the estimated time of arrival at the airport, the weather reports or forecasts or any combination of them, indicate:

    (i) The ceiling will be at least 2,000 feet above the airport elevation; and

    (ii) Visibility will be at least 3 miles.

91.25 VOR Equipment Check for IFR Operations

(a) No person may operate a civil aircraft under IFR using the VOR system of radio navigation unless the VOR equipment of that aircraft:

(1) Is maintained, checked, and inspected under an approved procedure; or

(2) Has been operationally checked within the preceding 30 days and was found to be within the limits of the permissible indicated bearing error set forth in paragraph (b) or (c) of this section.

(b) Except as provided in paragraph (c) of this section, each person conducting a VOR check under subparagraph (a) (2) of this section, shall:

(1) Use, at the airport of intended departure, an FAA operated or approved test signal or a test signal radiated by a certificated and appropriately rated radio repair station, or outside the United States, a test signal operated or approved by appropriate authority, to check the VOR equipment (the maximum permissible indicated bearing error is plus or minus 4 degrees);

(2) If a test signal is not available at the airport of intended departure, use a point on an airport surface designated as a VOR system checkpoint by the Administrator or, outside the United States, by appropriate authority (the maximum permissible bearing error is plus or minus 4 degrees);

(3) If neither a test signal nor a designated checkpoint on the surface is available, use an airborne checkpoint designated by the Administrator, or outside the United States, by appropriate authority (the maximum permissible bearing error is plus or minus 6 degrees); or

(4) If no check signal or point is available, while in flight:

    (i) Select a VOR radial that lies along the centerline of an established VOR airway;

    (ii) Select a prominent ground point along the selected radial preferably more than 20 miles from the VOR ground facility and maneuver the aircraft directly over the point at a reasonably low altitude; and

    (iii) Note the VOR bearing indicated by the receiver when over the ground point (the maximum permissible variation between the published radial and the indicated bearing is 6 degrees).

(c) If dual system VOR (units independent of each other except for the antenna) is installed in the aircraft, the person checking the equipment may check one system against the other in place of the check procedures specified in paragraph (b) of this section. He shall tune both systems to the same VOR ground facility and note the indicated bearings to that station. The maximum permissible variation between the two indicated bearings is 4 degrees.

(d) Each person making the VOR operational check as specified in paragraph (b) or (c) of this section shall enter the date, place, bearing error, and sign the aircraft log or other record. In addition, if a test signal radiated by a repair station, as specified in paragraph (b) (1) of this section, is used, an entry must be made in the aircraft log or other record by the repair station certificate holder or the certificate holder's representative certifying to the bearing transmitted by the repair station for the check and the date of transmission.

91.33 Powered Civil Aircraft with Standard Category U.S. Airworthiness Certificates; Instruments and Equipment

(b) *Visual flight rules (day).* For VFR flight during the day the following instruments and equipment are required:

(1) Airspeed indicator.

(2) Altimeter.

(3) Magnetic direction indicator.

(4) Tachometer for each engine.

(5) Oil pressure gauge for each engine using pressure system.

(6) Temperature gauge for each liquid-cooled engine.

(7) Oil temperature gauge for each air-cooled engine.

(8) Manifold pressure gauge for each altitude engine.

(9) Fuel gauge indicating the quantity of fuel in each tank.

(10) Landing gear position indicator, if the aircraft has a retractable landing gear.

(c) *Visual flight rules (night).* For VFR flight at night the following instruments and equipment are required:

(1) Instruments and equipment specified in paragraph (b) of this section.

(2) Approved position lights.

(3) An approved aviation red or aviation white anticollision light system on all U.S. registered civil aircraft.

(4) If the aircraft is operated for hire, one electric landing light.

(5) An adequate source of electrical energy for all installed electrical and radio equipment.

(6) One spare set of fuses, or three spare fuses of each kind required.

(d) *Instrument flight rules.* For IFR flight the following instruments and equipment are required:

(1) Instruments and equipment specified in paragraph (b) of this section and for night flight, instruments and equipment specified in paragraph (c) of this section.

(2) Two-way radio communications system and navigational equipment appropriate to the ground facilities to be used.

(3) Gyroscopic rate-of-turn indicator.

(4) Slip-skid indicator.

(5) Sensitive altimeter adjustable for barometric pressure.

(6) A clock displaying hours, minutes, and seconds with a sweep-second pointer or digital presentation.

(7) Generator of adequate capacity.

(8) Gyroscopic bank and pitch indicator (artificial horizon).

(9) Gyroscopic direction indicator (directional gyro or equivalent).

(e) *Flight at and above 24,000 feet MSL.* If VOR navigational equipment is required under paragraph (d) (2) of this section, no person may operate a U.S.registered civil aircraft within the 50 states, and the District of Columbia, at or above 24,000 feet MSL unless that aircraft is equipped with approved distance measuring equipment (DME).

91.75 Compliance with ATC Clearances and Instructions

(a) When an ATC clearance has been obtained, no pilot in command may deviate from that clearance, except in an emergency, unless he obtains an amended clearance. However, except in positive controlled airspace, this paragraph does not prohibit him from cancelling an IFR flight plan if he is operating in VFR weather conditions. If a pilot is uncertain of the meaning of an ATC clearance, he shall immediately request clarification from ATC.

(b) Except in an emergency, no person may in an area in which air traffic control is exercised, operate an aircraft contrary to an ATC instruction.

(c) Each pilot in command who deviates, in an emergency, from an ATC clearance or instruction shall notify ATC of that deviation as soon as possible.

(d) Each pilot in command who (though not deviating from a rule of this subpart) is given priority by ATC in an emergency, shall, if requested by ATC, submit a detailed report of that emergency within 48 hours to the chief of that ATC facility.

91.97 Positive Control Areas and Route Segments

(a) Except as provided in paragraph (b) of this section, no person may operate an aircraft within a positive control area or positive control route segment, designated in Part 71 of this chapter, unless that aircraft is:

(1) Operated under IFR at a specific flight level assigned by ATC;

(2) Equipped with instruments and equipment required for IFR operations;

(3) Flown by a pilot rated for instrument flight; and

(4) Equipped, when in a positive control area, with the applicable equipment specified in 91.24 and a radio providing direct pilot controller communication on the frequency specified by ATC for the area concerned.

91.107 Special VFR Weather Minimums

(e) No person may operate an aircraft (other than a helicopter) in a control zone under the special weather minimums of this section, between sunset and sunrise (or in Alaska, when the sun is more than six degrees below the horizon) unless:

    (1)That person meets the applicable requirements for instrument flight under Part 61 of this chapter; and

    (2)The aircraft is equipped as required in 91.33(d).

INSTRUMENT FLIGHT RULES

91.115 ATC Clearance and Flight Plan Required

No person may operate an aircraft in controlled airspace under IFR unless;

(a) He has filed an IFR flight plan; and

(b) He has received an appropriate ATC clearance.

91.116 Takeoff and Landing Under IFR

(a) *Instrument approaches to civil airports.* Unless otherwise authorized by the Administrator for paragraphs (a) through (k) of this section, when an instrument letdown to a civil airport is necessary, each person operating an aircraft, except a military aircraft of the United States, shall use a standard instrument approach procedure prescribed for the airport in Part 97 of this chapter.

(b) *Authorized DH or MDA.* For the purpose of this section, when the approach procedure being used provides for and requires use of a DH or MDA, the authorized decision height or authorized minimum descent altitude is the DH or MDA prescribed by the approach procedure, the DH or MDA prescribed for the pilot in command, or the DH or MDA for which the aircraft is equipped, whichever is higher.

(c) *Operation below DH or MDA.* Where a DH or MDA is applicable, no pilot may operate an aircraft, except a military aircraft of the United States, at any airport below the authorized MDA or continue an approach below the authorized DH unless:

    (1) The aircraft is continuously in a position from which a descent to a landing on the intended runway can be made at a normal rate of descent using normal maneuvers, and for operations conducted under Part 121 or Part135 unless that descent rate will allow touchdown to occur within the touchdown zone of the runway of intended landing;

    (2) The flight visibility is not less than the visibility prescribed in the standard instrument approach being used;

    (3) Except for a Category II or Category III approach where necessary visual reference requirements are specified by the Administrator, at least one of the following visual references for the intended runway is distinctly visible and identifiable to the pilot:

    (i) The approach light system, except that the pilot may not descend below 100 feet above the touchdown zone elevation using the approach lights as a reference unless the red terminating bars or the red side row bars are also distinctly visible and identifiable.

    (ii) The threshold.

    (iii) The threshold markings.

    (iv) The threshold lights.

    (v) The runway end identifier lights.

    (vi) The visual approach slope indicator.

    (vii) The touchdown zone or touchdown zone markings.

    (viii) The touchdown zone lights.

    (ix) The runway or runway markings.

    (x) The runway lights; and

(4) When the aircraft is on a straight-in nonprecision approach procedure which incorporates a visual descent point, the aircraft has reached the visual descent point, except where the aircraft is not equipped for or capable of establishing the point or a descent to the runway cannot be made using normal procedures or rates of descent if descent is delayed until reaching the point.

(d) *Landing.* No pilot operating an aircraft, except a military aircraft of the United States, may land that aircraft when the flight visibility is less than the visibility prescribed in the standard instrument approach procedure being used.

(e) *Missed approach procedures.* Each pilot operating an aircraft, except a military aircraft of the United States, shall immediately execute an appropriate missed approach procedure when either of the following conditions exist:

(1) Whenever the requirements of paragraph (c) of this section are not met at either of the following times:

    (i) When the aircraft is being operated below MDA;or

    (ii) Upon arrival at the missed approach point, including a DH where a DH is specified and its use is required, and at any time after that until touchdown.

(2) Whenever an identifiable part of the airport is not distinctly visible to the pilot during a circling maneuver at or above MDA, unless the inability to see an identifiable part of the airport results only from a normal bank of the aircraft during the circling approach.

(f) *Civil airport takeoff minimums.* Unless otherwise authorized by the Administrator, no pilot operating an aircraft under Part 121, 123, 125, 129, or 135 of this chapter may take off from a civil airport under IFR unless weather conditions are at or above the weather minimums for IFR takeoff prescribed for that airport under Part 97 of this chapter. If takeoff minimums are not prescribed under Part 97 of this chapter for a particular airport, the following minimums apply to takeoffs under IFR for aircraft operating under those Parts:

(1) For aircraft having two engines or less—1 statute mile visibility.

(2) For aircraft having more than two engines—½ statute mile visibility.

(i) *Operations on unpublished routes and use of radar in instrument approach procedures.* When radar is approved at certain locations for ATC purposes, it may be used not only for surveillance and precision radar approaches, as applicable, but also may be used in conjunction with instrument approach procedures predicted on other types of radio navigational aids. Radar vectors may be authorized to provide course guidance through the segments of an approach to the final course or fix. When operating on an unpublished route or while being radar vectored, the pilot, when an approach clearance is received, shall, in addition to complying with 91.119, maintain the last altitude assigned to that pilot until the aircraft is established on a segment of a published route or instrument approach procedure unless a different altitude is assigned by ATC. After the aircraft is so established, published altitudes apply to descent within each succeeding route or approach segment unless a different altitude is assigned by ATC. Upon reaching the final approach course or fix, the pilot may either complete the instrument approach in accordance with a procedure approved for the facility or continue a surveillance or precision radar approach to a landing.

(j) *Limitation on procedure turns.* In the case of a radar vector to a final approach course or fix, a timed approach from a holding fix, or an approach for which the procedure specifies "No PT", no pilot may make a procedure turn unless cleared to do so by ATC.

(k) *ILS components.* The basic ground components of an ILS are the localizer, glide slope, outer marker, middle marker, and, when installed for use with Category II or Category III instrument approach procedures and inner marker. A compass locator or precision radar may be substituted for the outer or middle marker. DME, VOR, or nondirectional beacon fixes authorized in the standard instrument procedure or surveillance radar may be substituted for the outer marker. Applicability of, and substitution for, the inner marker for Category II or III approaches is determined by the appropriate Part 97 approach procedure, letter of authorization, or operations specification pertinent to the operations.

91.119 Minimum Altitudes for IFR Operations

(a) Except when necessary for takeoff or landing, or unless otherwise authorized by the Administrator, no person may operate an aircraft under IFR below:

    (1) The applicable minimum altitudes prescribed in Parts 95 and 97 of this chapter; or

    (2) If no applicable minimum altitude is prescribed in those Parts:

        (i) In the case of operations over an area designated as a mountainous area in Part 95, an altitude of 2,000 feet above the highest obstacle within a horizontal distance of five statute miles from the course to be flown; or

        (ii) In any other case, an altitude of 1,000 feet above the highest obstacle within a horizontal distance of five statute miles from the course to be flown. However, if both a MEA and a MOCA are prescribed for a particular route or route segment, a person may operate an aircraft below the MEA down to, but not below, the MOCA, when within

25 statute miles of the VOR concerned (based on the pilot's reasonable estimate of that distance).

(b) *Climb.* Climb to a higher minimum IFR altitude shall begin immediately after passing the point beyond which that minimum altitude applies, except that, when ground obstructions intervene, the point beyond which the higher minimum altitude applies shall be crossed at or above the applicable MCA.

91.121 IFR Cruising Altitude or Flight Level

(a) *In controlled airspace.* Each person operating an aircraft under IFR in level cruising flight in controlled airspace shall maintain the altitude or flight level assigned that aircraft by ATC. However, if the ATC clearance assigns "VFR conditions-on-top," he shall maintain altitude or flight level as prescribed by 91.109.

(b) *In uncontrolled airspace.* Except while holding in a holding pattern of two minutes or less, or while turning, each person operating an aircraft under IFR in level cruising flight, in uncontrolled airspace, shall maintain an appropriate altitude as follows:

    (1) When operating below 18,000 feet MSL and

        (i) On a magnetic course of zero degrees through 179 degrees, any odd thousand foot MSL altitude (such as 3,000, 5,000, or 7,000; or

        (ii) On a magnetic course of 180 degrees through 359 degrees, any even thousand foot MSL altitude (such as 2,000, 4,000, or 6,000).

91.123 Course To Be Flown

Unless otherwise authorized by ATC, no person may operate an aircraft within controlled airspace, under IFR, except as follows:

(a) On a Federal airway, along the centerline of that airway.

(b) On any other route, along the direct course between the navigational aids or fixes defining that route. However, this section does not prohibit maneuvering the aircraft to pass well clear of other air traffic or the maneuvering of the aircraft, in VFR conditions, to clear the intended flight path both before and during climb or descent.

91.125 IFR Radio Communications

The pilot in command of each aircraft operated under IFR in controlled airspace shall have a continuous watch maintained on the appropriate frequency and shall report by radio as soon as possible:

(a) The time and altitude of passing each designated reporting point, or the reporting points specified by ATC, except that while the aircraft is under radar control, only the passing of those reporting points specifically requested by ATC need be reported;

(b) Any unforecast weather conditions encountered; and

(c) Any other information relating to the safety of flight.

91.127 IFR Operations: Two-Way Radio Communications Failure

(a) *General.* Unless otherwise authorized by ATC, each pilot who has two-way radio communications failure when operating under IFR shall comply with the rules of this section.

(b) *VFR conditions.* If the failure occurs in VFR conditions, or if VFR conditions are encountered after the failure, each pilot shall continue the flight under VFR and land as soon as practicable.

(c) *IFR conditions.* If the failure occurs in IFR conditions, or if paragraph (b) of this section cannot be complied with, each pilot shall continue the flight according to the following:

    (1) *Route.*

        (i) By the route assigned in the last ATC clearance received;

        (ii) If being radar vectored, by the direct route from the point of radio failure to the fix, route, or airway specified in the vector clearance;

        (iii) In the absence of an assigned route, by the route that ATC has advised may be expected in a further clearance; or

        (iv) In the absence of an assigned route or a route that ATC has advised may be expected in a further clearance, by the route filed in the flight plan.

    (2) *Altitude.* At the highest of the following altitudes or flight levels for the route segment being flown:

        (i) The altitude or flight level assigned in the last ATC clearance received;

        (ii) The minimum altitude (converted, if appropriate, to minimum flight level as prescribed in 91.81 (c) for IFR operations; or

        (iii) The altitude or flight level ATC has advised may be expected in a further clearance.

    (3) *Leave clearance limit.*

        (i) When the clearance limit is a fix from which an approach begins, commence descent or descent and approach as close as possible to the expect further clearance time if one has been received, or if one has not been received, as close as possible to the estimated time of arrival as calculated from the filed or amended (with ATC) estimated time en route.

        (ii) If the clearance limit is not a fix from which an approach begins, leave the clearance limit at the expect further clearance time if one has been received, or if none has been received, upon arrival over the clearance limit, and proceed to a fix from which an approach begins and commence descent or descent and approach as close as possible to the estimated time of arrival as calculated from the filed or amended (with ATC) estimated time en route.

91.129 Operation Under IFR In Controlled Airspace; Malfunction Reports

(a) The pilot in command of each aircraft operated in controlled airspace under IFR, shall report immediately to ATC any of the following malfunctions of equipment occurring:

   (1) Loss of VOR, TACAN, ADF, or low frequency navigation receiver capability.

   (2) Complete or partial loss of ILS receiver capability.

   (3) Impairment of air/ground communications capability.

(b) In each report required by paragraph (a) of this section, the pilot in command shall include the:

   (1) Aircraft identification;

   (2) Equipment affected;

   (3) Degree to which the capability of the pilot to operate under IFR in the ATC system is impaired; and

   (4) Nature and extent of assistance he desires from ATC.

Glossary

The terms in this glossary are excerpted from the "Pilot/Controller Glossary" that appears at the end of each edition of the *Airman's Information Manual* (AIM).

The glossary in AIM was compiled by the FAA to promote a common understanding of the terms used in the air traffic control system. It includes those terms that are intended for pilot/controller communications.

Be familiar with all the terms and definitions in the complete glossary. However, you must know the definitions excerpted and reprinted here in order to be a competent instrument pilot.

Abbreviated IFR Flight Plans—An authorization by ATC requiring pilots to submit only that information needed for the purpose of ATC. It includes only a small portion of the usual IFR flight plan information. In certain instances, this may be only aircraft identification, location, and pilot request. Other information may be requested if needed by ATC for separation/control purposes. It is frequently used by aircraft which are airborne and desire an instrument approach or by aircraft which are on the ground and desire a climb to VFR-on-top. (Refer to AIM).

Advisory Frequency—The appropriate frequency to be used for Airport Advisory Service. (Refer to Advisory Circular No. 90-42D and AIM).

affirmative—Yes.

airport advisory area—The area within 10 miles of an airport without a control tower or where the tower is not in operation, and on which a flight service station is located. (Refer to AIM.)

airport advisory service/AAS—A service provided by flight service stations or the military at airports not serviced by an operating control tower. This service consists of providing information to arriving and departing aircraft concerning wind direction and speed, favored runway, altimeter setting, pertinent known traffic, pertinent known field conditions, airport taxi routes and traffic patterns, and authorized instrument approach procedures. This information is advisory in nature and does not constitute an ATC clearance.

air route surveillance radar/ARSR—Air route traffic control center (ARTCC) radar used primarily to detect and display an aircraft's position while en route between terminal areas. The ARSR enables controllers to provide radar air traffic control service when aircraft are within the ARSR coverage. In some instances, ARSR may enable an ARTCC to provide terminal radar services similar to but usually more limited than those provided by a radar approach control.

air route traffic control center/ARTCC—A facility established to provide air traffic control service to aircraft operating on IFR flight plans within controlled airspace and principally during the en route phase of flight. When equipment capabilities and controller workload permit, certain advisory/assistance services may be provided to VFR aircraft. (Refer to AIM.)

air traffic control/ATC—A service operated by appropriate authority to promote the safe, orderly and expeditious flow of air traffic.

alternate airport—An airport at which an aircraft may land if a landing at the intended airport becomes inadvisable.

approach clearance—Authorization by ATC for a pilot to conduct an instrument approach. The type of instrument approach for which a clearance and other pertinent information is provided in the approach clearance when required. (Refer to AIM and FAR Part 91.)

approach control facility—A terminal ATC facility that provides approach control service in a terminal area.

area navigation/RNAV—A method of navigation that permits aircraft operation on any desired course within the coverage of station-referenced navigation signals or within the limits of a self-contained system capability. Random area navigation routes are direct routes, based on area navigation capability, between waypoints defined in terms of latitude/longitude coordinates, degree/distance fixes, or offsets from published or established routes/airways at a specified distance and direction. The major types of equipment are:

- VORTAC referenced or course line computer (CLC) systems, which account for the greatest number of RNAV units in use. To function, the CLC must be within the service range of a VORTAC.
- OMEGA/VLF, although two separate systems, can be considered as one operationally. A long-range navigation system based upon very low frequency radio signals transmitted from a total of 17 stations worldwide.
- Inertial (INS) systems, which are totally self-contained and require no information from external references. They provide aircraft position and navigation information in response to signals resulting from inertial effects on components within the system.

- MLS area navigation (MLS/RNAV), which provides area navigation with reference to an MLS ground facility.
- LORAN-C is a long-range navigation system that used ground waves transmitted at low frequency to provide user position information at ranges of up to 600 to 1,200 nautical miles at both en route and approach altitudes. The useable signal coverage areas are determined by the signal-to-noise ratio, the envelope-to -cycle difference, and the geometric relationship between the positions of the user and the transmitting stations.

ATC clears—Used to prefix an ATC clearance when it is relayed to an aircraft by other than an air traffic controller.

ATC instructions—Directives issued by air traffic control for the purpose of requiring a pilot to take specific actions; e.g., ''Turn left heading two five zero,'' ''Go around,'' ''Clear the runway.'' (Refer to FAR Part 91.)

automatic altitude reporting—That function of a transponder which responds to Mode C interrogations by transmitting the aircraft's altitude in 100-foot increments.

automatic direction finder/ADF—An aircraft radio navigation system which senses and indicates the direction to a L/MF nondirectional radio beacon (NDB) ground transmitter. Direction is indicated to the pilot as a magnetic bearing or as a relative bearing to the longitudinal axis of the aircraft depending on the type of indicator installed in the aircraft. In certain applications, such as military, ADF operations may be based on airborne and ground transmitters in the VHF/UHF frequency spectrum.

automatic terminal information service/ATIS—The continuous broadcast of recorded noncontrol information in selected terminal areas. Its purpose is to improve controller effectiveness and to relieve frequency congestion by automating the repetitive transmission of essential but routine information; e.g., ''Los Angeles information Alfa. One three zero zero Coordinated Universal Time. Weather, measured ceiling two thousand overcast, visibility three, haze, smoke, temperature seven one, dew point five seven, wind two five zero at five, altimeter two niner niner six. I-L-S Runway Two Five Left approach in use, Runway Two Five Right closed, advise you have Alfa.'' (Refer to AIM.)

aviation weather service—A service provided by the National Weather Service (NWS) and FAA which collects and disseminates pertinent weather information for pilots, aircraft operators, and ATC. Available aviation weather reports and forecasts are displayed at each NWS office and FAA FSS. (Refer to AIM.)

center weather advisory/CWA—An unscheduled weather advisory issued by Center Weather Service Unit meteorologists for ATC use to alert pilots of existing or anticipated adverse weather conditions within the next 2 hours. A CWA may modify or redefine a SIGMET. (Refer to AIM.)

circle-to-land maneuver/circling maneuver—a maneuver initiated by the pilot to align the aircraft with a runway for landing when a straight-in landing from an instrument approach is not possible or is not desirable. This maneuver is made only after ATC authorization has been obtained and the pilot has established required visual reference to the airport. (Refer to AIM.)

clearance limit—The fix, point, or location to which an aircraft is cleared when issued an air traffic clearance.

clearance void if not off by (time)—Used by ATC to advise an aircraft that the departure clearance is automatically cancelled if takeoff is not made prior to a specified time. The pilot must obtain a new clearance or cancel his IFR flight plan if not off by the specified time.

cleared as filed—Means the aircraft is cleared to proceed in accordance with the route of flight filed in the flight plan. This clearance does not include the altitude, SID, or SID Transition. (Refer to AIM.)

cleared for (type of) approach—ATC authorization for an aircraft to execute a specific instrument approach procedure to an airport; e.g., ''Cleared for ILS Runway Three Six Approach.'' (Refer to AIM, FAR Part 91.)

cleared for approach—ATC authorization for an aircraft to execute any standard or special instrument approach procedure for that airport. Normally, an aircraft will be cleared for a specific instrument approach procedure. (Refer to AIM, FAR Part 91.)

codes/transponder codes—The number assigned to a particular multiple pulse reply signal transmitted by a transponder.

common transponder codes—The number assigned to a particular multiple pulse reply signal transmitted by a transponder.

common traffic advisory frequency/CTAF—A frequency designed for the purpose of carrying out airport advisory practices while operating to or from an uncontrolled airport. The CTAF may be a UNICOM, Multicom, FSS, or tower frequency and is identified in appropriate aeronautical publications. (Refer to AC—90-42C.)

compass locator—A low power, low or medium frequency (L/MF) radio beacon installed at the site of the outer or middle marker of an instrument landing system (ILS). It can be used for navigation at distances of approximately 15 miles or as authorized in the approach procedure.

- Outer Compass Locator/LOM—A compass locator installed at the site of the outer marker of an instrument landing system.
- Middle Compass Locator/LMM—A compass locator installed at the site of the middle marker of an instrument landing system.

contact approach—An approach wherein an aircraft on an IFR flight plan, having an air traffic control authorization, operating clear of clouds with at least 1 mile flight visibility and a reasonable expectation of continuing to the destination airport in those conditions, may deviate from the instrument approach procedure and proceed to the destination airport by visual reference to the surface. This approach will only be authorized when requested by the pilot and the reported ground visibility at the destination airport is at least 1 statute mile. (Refer to AIM.)

Convective SIGMET/WST/convective significant meteorological information—A weather advisory concerning convective weather significant to the safety of all aircraft.

Convective SIGMETs are issued for tornadoes, lines of thunderstorms, embedded thunderstorms of any intensity level, areas of thunderstorms greater than or equal to VIP level 4 with an aerial coverage of $\frac{4}{10}$ (40 percent) or more, and hail $\frac{3}{4}$ inch or greater. (Refer to AIM.)

cruise—Used in an ATC clearance to authorize a pilot to conduct flight at any altitude from the minimum IFR altitude up to and including the altitude specified in the clearance. The pilot may level off at any intermediate altitude within this block of airspace. Climb/descent within the block is to be made at the discretion of the pilot. However, once the pilot starts descent and verbally reports leaving an altitude in the block, he may not return to that altitude without additional ATC clearance. Further, it is approval for the pilot to proceed to and make an approach at destination airport.

decision height/DH—With respect to the operation of aircraft, means the height at which a decision must be made during an ILS, MLS, or PAR instrument approach to either continue the approach or to execute a missed approach.

delay indefinite (reason if known) expect further clearance time—Used by ATC to inform a pilot when an accurate estimate of the delay time and the reason for the delay cannot immediately be determined; e.g., a disabled aircraft on the runway, terminal or center area saturation, weather below landing minimums, etc.

departure control—A function of an approach control facility providing air traffic control service for departing IFR and, under certain conditions, VFR aircraft. (Refer to AIM.)

deviations—

- A departure from a current clearance, such as an off course maneuver to avoid weather or turbulence.
- Where specifically authorized in the FARs and requested by the pilot, ATC may permit pilots to so deviate from certain regulations. (Refer to AIM.)

displaced threshold—A threshold that is located at a point on the runway other than the designated beginning of the runway.

distance measuring equipment/DME—Equipment (airborne and ground) used to measure, in nautical miles, the slant range distance of an aircraft from the DME navigational aid.

emergency locator transmitter/ELT—A radio transmitter attached to the aircraft structure which operates from its own power source on 121.5 MHz and 243.0 MHz. It aids in locating downed aircraft by radiating a downward sweeping audio tone, 2-4 times per second. It is designed to function without human action after an accident. (Refer to FAR, Part 91, AIM.)

en route air traffic control services—Air traffic control service provided aircraft on IFR flight plans, generally by centers, when these aircraft are operating between departure and destination terminal areas. When equipment, capabilities, and controller workload permit, certain advisory/assistance services may be provided to VFR aircraft. (Refer to AIM.)

en route flight advisory service/FLIGHT WATCH—A weather service specifically designed to provide, upon pilot request, timely weather information pertinent to his type of flight, intended route of flight, and altitude. The FSSs providing this service are listed in the Airport/Facility Directory. (Refer to AIM.)

expect (altitude) at (time) or (fix)—Used under certain conditions to provide a pilot with an altitude to be used in the event of two-way communications failure. It also provides altitude information to assist the pilot in planning. (Refer to AIM.)

expected departure clearance time/EDCT—The runway release time assigned to an aircraft in a controlled departure time program and shown on the flight progress strip as an EDCT.

expect further clearance (time)/EFC—The time a pilot can expect to receive clearance beyond a clearance limit.

final approach fix/FAF—The fix from which the final approach (IFR) to an airport is executed and which identifies the beginning of the final approach segment. It is designated on government charts by the Maltese Cross symbol for nonprecision approaches and the lightning bolt symbol for precision approaches; or when ATC directs a lower-than-published glideslope/path intercept altitude, it is the resultant actual point of the glideslope/path intercept.

final approach-IFR—The flight path of an aircraft which is inbound to an airport on a final instrument approach course, beginning at the final approach fix or point and extending to the airport or the point where a circle-to-land maneuver or a missed approach is executed.

final service station/FSS—Air traffic facilities which provide pilot briefing, en route communications and VFR search and rescue services, assist lost aircraft and aircraft in emergency situations, relay ATC clearances, originate Notices to Airmen, broadcast aviation weather and NAS information, receive and process IFR flight plans, and monitor NAVAIDs. In addition, at selected locations, FSSs provide en route flight advisory service (Flight Watch), take weather observations, issue airport advisories, and advise customs and immigration of transborder flights. (Refer to AIM.)

flight standards district office/FSDO—An FAA field office serving an assigned geographical area and staffed with flight standards personnel who serve the aviation industry and the general public on matters relating to the certification and operation of air carrier and general aviation aircraft. Activities include general surveillance of operational safety, certification of airmen and aircraft, accident prevention, investigation, enforcement, etc.

flight test—A flight for the purpose of:

- Investigating the operation/flight characteristics of an aircraft or aircraft component.
- Evaluating an applicant for a pilot certificate or rating.

gate hold procedures—Procedures at selected airports to hold aircraft at the gate or other ground location whenever departure delays exceed or are anticipated to exceed 15 minutes. The sequence for departure will be maintained in accordance with initial call-up unless modified by flow control restrictions. Pilots should monitor the ground control/clearance delivery frequency for engine startup advisories or new proposed start time if the delay changes.

glideslope/glidepath—Provides vertical guidance for aircraft during approach and landing. The glideslope/glidepath is based on the following:

1. Electronic components emitting signals which provide vertical guidance by reference to airborne instruments during instrument approaches such as ILS/MLS.
2. Visual ground aids, such as VASI, which provide vertical guidance for a VFR approach or for the visual portion of an instrument approach and landing.
3. PAR. Used by ATC to inform an aircraft making a PAR approach of its vertical position (elevation) relative to the descent profile.

glideslope/glidepath intercept altitude—The minimum altitude to intercept the glideslope/path on a precision approach. The intersection of the published intercept altitude with the glideslope/path, designated on Government charts by the lightning bolt symbol, is the precision FAF; however, when ATC directs a lower altitude, the resultant lower intercept position is then the FAF.

ground delay—The amount of delay attributed to ATC, encountered prior to departure, usually associated with a CDT program.

ground speed—The speed of an aircraft relative to the surface of the earth.

handoff—An action taken to transfer the radar identification of an aircraft from one controller to another if the aircraft will enter the receiving controller's airspace and radio communications with the aircraft will be transferred.

height above airport/HAA—The height of the Minimum Descent Altitude above the published airport elevation. This is published in conjunction with circling minimums.

height above landing/HAL—The height above a designated helicopter landing area used for helicopter instrument approach procedures. (Refer to FAR Part 97.)

height above touchdown/HAT—The height of the decision height or minimum descent altitude above the highest runway elevation in the touchdown zone (first 3,000 feet of the runway). HAT is published on instrument approach charts in conjunction with all straight-in minimums. (See decision height, minimum descent altitude.)

hold/holding procedure—A predetermined maneuver which keeps aircraft within a specified airspace while awaiting further clearance from air traffic control. Also used during ground operations to keep aircraft within a specified area or at a specified point while awaiting further clearance from air traffic control. (Refer to AIM.)

holding fix—A specified fix identifiable to a pilot by NAVAIDs or visual reference to the ground used as a reference point in establishing and maintaining the position of an aircraft while holding. (Refer to AIM.)

hold for release—Used by ATC to delay an aircraft for traffic management reasons; i.e., weather, traffic volume. etc. Hold for release instructions (including departure delay information) are used to inform a pilot or a controller (either directly or through an authorized relay) that a departure clearance is not valid until a release time or additional instructions have been received.

homing—Flight toward a navaid, without correcting for wind, by adjusting the aircraft heading to maintain a relative bearing of zero degrees.

IFR aircraft/IFR flight—An aircraft conducting flight in accordance with instrument flight rules.

IFR conditions—Weather conditions below the minimum for flight under visual flight rules.

IFR takeoff minimums and departure procedures—FAR, Part 91, prescribes standard takeoff rules for certain civil users. At some airports, obstructions or other factors require the establishment of nonstandard takeoff minimums, departure procedures, or both to assist pilots in avoiding obstacles during climb to the minimum en route altitude. Those airports are listed in NOS/DOD instrument approach charts (IAPs) under a section entitled "IFR Takeof Minimums and Departure Procedures." The NOS/DOD IAP chart legend illustrates the symbol used to alert the pilot to nonstandard takeoff minimums and departure procedures. When departing IFR from such airports or from any airports where there are no departure procedures, SIDs, or ATC facilities available, pilots should advise ATC of any departure limitations. Controllers may query a pilot to determine acceptable departure directions, turns, or headings after takeoff. Pilots should be familiar with the departure procedures and must assure that their aircraft can meet or exceed any specified climb gradients.

ILS Categories—

ILS Category I—An ILS approach procedure which provides for approach to a height above touchdown of not less than 200 feet and with runway visual range of not less than 1,800 feet.

ILS Category II—An ILS approach procedure which provides for approach to a height above touchdown of not less than 100 feet and with runway visual range of not less than 1,200 feet.

ILS Category III

a. IIIA—An ILS approach procedure which provides for approach without a decision height minimum and with runway visual range of not less than 700 feet.

b. IIIB—An ILS approach procedure which provides for approach without a decision height minimum and with runway visual range of not less than 150 feet.

c. IIIC—An ILS approach procedure which provides for approach without a decision height minimum and without runway visual range minimum.

initial approach fix/IAF—The fixes depicted on instrument approach procedure charts that identify the beginning of the initial approach segment(s).

instrument approach procedure/IAP/instrument approach—A series of predetermined maneuvers for the orderly transfer of an aircraft under instrument flight conditions from the beginning of the initial approach to a landing or to a point from which a landing may be made visually. It is prescribed and approved for a specific airport by competent authority. (Refer to FAR, Part 91, AIM.)

1. U.S. civil standard instrument approach procedures are approved by the FAA as prescribed under FAR, Part 97 and are available for public use.

2. U.S. military standard instrument approach procedures are approved and published by the Department of Defense.
3. Special instrument approach procedures are approved by the FAA for individual operators but are not published in FAR, Part 97 for public use.

instrument flight rules/IFR—Rules governing the procedures for conducting instrument flight. Also a term used by pilots and controllers to indicate type of flight plan. (Refer to AIM.)

Instrument Landing System/ILS—A precision instrument approach system which normally consists of the following electronic components and visual aids:

- localizer. (See localizer.)
- glideslope. (See glideslope.)
- outer marker. (See outer marker.)
- middle marker. (See middle marker.)
- approach lights. (See airport lighting.)

(Refer to FAR Part 91, AIM.)

instrument meteorological conditions/IMC—Meteorological conditions expressed in terms of visibility, distance from cloud, and ceiling less than the minima specified for visual meteorological conditions.

instrument runway—A runway equipped with electronic and visual navigation aids for which a precision or nonprecision approach procedure having straight-in landing minimums has been approved.

intersection—A point defined by any combination of courses, radials, or bearings of two or more navigational aids; used to describe the point where two runways, a runway and a taxiway, or two taxiways cross or meet.

jet route—A route designed to serve aircraft operations from 18,000 feet MSL up to and including flight level 450. The routes are referred to as "J" routes with numbering to identify the designated route: e.g., J105. (Refer to FAR Part 71.)

jet stream—A migrating stream of high-speed winds present at high altitudes.

known traffic—With respect to ATC clearances, means aircraft whose altitude, position, and intentions are known to ATC.

landing minimums/IFR—The minimum visibility prescribed for landing a civil aircraft while using an instrument approach procedure. The minimum applies with other limitations set forth in FAR Part 91 with respect to the minimum descent altitude (MDA) or decision height (DH) prescribed in the instrument approach procedures as follows:

1. Straight-in landing minimums—A statement of MDS and visibility, or DH and visibility, required for a straight-in landing on a specified runway, or
2. Circling minimums—A statement of MDA and visibility required for the circle-to-land maneuver.

Descent below the established MDA or DH is not authorized during an approach unless the aircraft is in a position from which a normal approach to the runway of intended landing can be made and adequate visual reference to required visual cues is maintained. (Refer to FAR Part 91.)

localizer—The component of an ILS which provides course guidance to the runway.

localizer type directional aid/LDA—A NAVAID used for nonprecision instrument approaches with utility and accuracy comparable to a localizer but which is not part of a complete ILS and is not aligned with the runway. (Refer to AIM.)

localizer usable distance—The maximum distance from the localizer transmitter at a specified altitude, as verified by flight inspection, at which reliable course information is continuously received. (Refer to AIM.)

LORAN/long range navigation—An electronic navigational system by which hyperbolic lines of position are determined by measuring the difference in the time of reception of synchronized pulse signals from two fixed transmitters. Loran A operates in the 1750-1950 kHz frequency band. Loran C and D operate in the 100-110 kHz frequency band. (Refer to AIM.)

low altitude airway structure/federal airways—The network of airways serving aircraft operations up to but not including 18,000 feet MSL. (See airway.) (Refer to AIM.)

low altitude alert system/LAAS—An automated function of the TPX-42 that alerts the controller when a Mode C transponder-equipped aircraft on an IFR flight plan is below a predetermined minimum safe altitude. If requested by the pilot, LAAS monitoring is also available to VFR Mode C transponder-equipped aircraft.

low approach—An approach over an airport or runway following an instrument approach or a VFR approach including the go-around maneuver where the pilot intentionally does not make contact with the runway. (Refer to AIM.)

marker beacon—An electronic navigation facility transmitting a 75 MHz vertical fan or boneshaped radiation pattern. Marker beacons are identified by their modulation frequency and keying code, and when received by compatible airborne equipment, indicate to the pilot, both aurally and visually, that he is passing over the facility. (Refer to AIM.)

maximum authorized altitude/MAA—A published altitude representing the maximum usable altitude or flight level for an airspace structure or route segment. It is the highest altitude on a Federal airway, jet route, area navigation low or high route, or other direct route for which an MEA is designated in FAR Part 95 at which adequate reception of navigation aid signals is assured.

middle marker/MM—A marker beacon that defines a point along the glide slope of an ILS normally located at or near the point of decision height (ILS Category I). It is keyed to transmit alternate dots and dashes, with the alternate dots and dashes keyed at the rate of 95 dot/dash combinations per minute on a 1300 Hz tone, which is received aurally and visually by compatible airborne equipment. (Refer to AIM.)

minimum crossing altitude/MCA—The lowest altitude at certain fixes at which an aircraft must cross when proceeding in the direction of a higher minimum en route IFR altitude (MEA).

minimum descent altitude/MDA—The lowest altitude, expressed in feet above mean sea level, to which descent is authorized on final approach or during circle-to-land maneuvering in execution of a standard instrument approach procedure where no electronic glide slope is provided.

minimum en route IFR altitude/MEA—The lowest published altitude between radio fixes which assures acceptable navigational signal coverage and meets obstacle clearance requirements between those fixes. The MEA prescribed for a Federal airway or segment thereof, area navigation low or high route, or other direct route applies to the entire width of the airway, segment, or route between the radio fixes defining the airway, segment, or route. (Refer to FAR Parts 91 and 95; AIM.)

minimum fuel—Indication that an aircraft's fuel supply has reached a state where, upon reaching the destination, it can accept little or no delay. This is not an emergency situation but merely indicates an emergency situation is possible should any undue delay occur. (Refer to AIM.)

minimum holding altitude/MHA—The lowest altitude prescribed for a holding pattern which assures navigational signal coverage, communications, and meets obstacle clearance requirements.

minimum IFR altitudes/MIA—Minimum altitudes for IFR operations as prescribed in FAR Part 91. These altitudes are published on aeronautical charts and prescribed in FAR Part 95 for airways and routes, and in FAR Part 97 for standard instrument approach procedures. If no applicable minimum altitude is prescribed in FAR Parts 95 or 97, the following minimum IFR altitude applies:

- In designated mountainous areas, 2,000 feet above the highest obstacle within a horizontal distance of 5 statute miles from the course to be flown; or
- Other than mountainous areas, 1,000 feet above the highest obstacle within a horizontal distance of 5 statute miles from the course to be flown; or
- As otherwise authorized by the Administrator or assigned by ATC. (Refer to FAR Part 91.)

minimum obstruction clearance altitude/MOCA—The lowest published altitude in effect between radio fixes on VOR airways, off-airway routes, or route segments which meets obstacle clearance requirements for the entire route segment and which assures acceptable navigational signal coverage only within 25 statute (22 nautical) miles of a VOR. (Refer to FAR Part 91 and 95.)

minimum reception altitude/MRA—The lowest altitude at which an intersection can be determined. (Refer to FAR Part 95.)

minimum safe altitude/MSA—1. The minimum altitude specified in FAR Part 91 for various aircraft operations. 2. Altitudes depicted on approach charts which provide at least 1,000 feet of obstacle clearance for emergency use within a specified distance from the navigation facility upon which a procedure is predicated. These altitudes will be identified as Minimum Sector Altitudes or Emergency Safe Altitudes and are established as follows:

a. Minimum Sector Altitudes—Altitudes depicted on approach charts which provide at least 1,000 feet of obstacle clearance within a 25-mile radius of the navigation facility upon which the procedure is predicated. Sectors depicted on approach charts must be at least 90 degrees in scope. These altitudes are for emergency use only and do not necessarily assure acceptable navigational signal coverage.

minimum safe altitude warning/MSAW—A function of the ARTS III computer that aids the controller by alerting him when a tracked Mode C- equipped aircraft is below or is predicted by the computer to go below a predetermined minimum safe altitude. (Refer to AIM.)

minimums/minima—Weather condition requirements established for a particular operation or type of operation; e.g., IFR takeoff or landing, alternate airport for IFR flight plans, VFR flight, etc. (Refer to FAR Part 91, AIM.)

minimum vectoring altitude/MVA—The lowest MSL altitude at which an IFR aircraft will be vectored by a radar controller, except as otherwise authorized for radar approaches, departures, and missed approaches. The altitude meets IFR obstacle clearance criteria. It may be lower than the published MEA along an airway or J-route segment. It may be utilized for radar vectoring only upon the controller's determination that an adequate radar return is being received from the aircraft being controlled. Charts depicting minimum vectoring altitudes are normally available only to the controllers and not to pilots. (Refer to AIM.)

missed approach—

- A maneuver conducted by a pilot when an instrument approach cannot be completed to a landing. The route of flight and altitude are shown on instrument approach procedure charts. A pilot executing a missed approach prior to the missed approach point (MAP) must continue along the final approach to the MAP. The pilot may climb immediately to the altitude specified in the missed approach procedure.
- A term used by the pilot to inform ATC that he is executing the missed approach.
- At locations where ATC radar service is provided, the pilot should conform to radar vectors when provided by ATC in lieu of the published missed approach procedure. (See missed approach point.) (Refer to AIM.)

missed approach point/MAP—A point prescribed in each instrument approach procedure at which a missed approach procedure shall be executed if the required visual reference does not exist.

moving target indicator/MTI—An electronic device which will permit radar scope presentation only from targets which are in motion. A partial remedy for ground clutter.

National Airspace System/NAS—The common network of U.S. airspace; air navigation facilities, equipment and services, airports or landing areas; aeronautical charts, information and services; rules, regulations and procedures, technical information, and manpower and material. Included are system components shared jointly with the military.

navaid classes—VOR, VORTAC, and TACAN aids are classified according to their operational use. The three classes of NAVAIDs are: T—terminal, L—Low Altitude.

H—High Altitude. The normal service range for T, L, and H class aids is found in the AIM. Certain operational requirements make it necessary to use some of these aids at greater service ranges than specified. Extended range is made possible through flight inspection determinations. Some aids also have lesser service range due to location, terrain, frequency protection, etc. Restrictions to service range are listed in *Airport/Facility Directory.*

negative—No, or permission not granted, or that is not correct.

negative contact—Used by pilots to inform ATC that:

- Previously issued traffic is not in sight. It may be followed by the pilot's request for the controller to provide assistance in avoiding the traffic.
- They were unable to contact ATC on a particular frequency.

night—The time between the end of evening civil twilight and the beginning of morning civil twilight, as published in the *American Air Almanac*, converted to local time.

no gyro approach/vector—A radar approach/vector provided in case of a malfunctioning gyro-compass or directional gyro. Instead of providing the pilot with headings to be flown, the controller observes the radar track and issues control instructions "turn right/left" or "stop turn" as appropriate. (Refer to AIM.)

nondirectional beacon/radio beacon/NDB—An L/MF or UHF radio beacon transmitting nondirectional signals whereby the pilot of an aircraft equipped with direction finding equipment can determine his bearing to or from the radio beacon and home on or track to or from the station. When the radio beacon is installed in conjunction with the instrument landing system marker, it is normally called a compass locator.

nonprecision approach procedure/nonprecision approach—A standard instrument approach procedure in which no electronic glide slope is provided; e.g., VOR, TACAN, NDB, LOC, ASR, LDA, or SDF approaches.

nonradar—Precedes other terms and generally means without the use of radar, such as:

- Nonradar approach—Used to describe instrument approaches for which course guidance on final approach is not provided by ground-based precision or surveillance radar. Radar vectors to the final approach course may or may not be provided by ATC. Examples of nonradar approaches are VOR, NDB, TACAN, and ILS/MLS approaches.
- Nonradar approach control—An ATC facility providing approach control service without the use of radar.
- Nonradar arrival—An aircraft arriving at an airport without radar service, or at an airport served by a radar facility but without radar services being provided.
- Nonradar route—A flight path or route over which the pilot is performing his own navigation. The pilot may be receiving radar separation, radar monitoring, or other ATC services while on a nonradar route.
- Nonradar separation—The spacing of aircraft in accordance with established minima without the use of radar; e.g., vertical, lateral or longitudinal separation.

notices to airmen publication—A publication designed primarily as a pilot's operational manual containing current NOTAM information considered essential to the safety of flight as well as supplemental data to other aeronautical publications.

notice to airmen/NOTAM—A notice containing information (not known sufficiently in advance to publicize by other means) concerning the establishment, condition, or change in any component (facility, service, or procedure of, or hazard in the National Airspace System) the timely knowledge of which is essential to personnel concerned with flight operations.

- NOTAM(D)—A NOTAM given (in addition to local dissemination) distant dissemination beyond the area of responsibility of the Flight Service Station. These NOTAMs will be stored and available until canceled.
- NOTAM(L)—A NOTAM given local dissemination by voice and other means, such as teleautograph and telephone, to satisfy local user requirements.
- FDC NOTAM—A NOTAM regulatory in nature, transmitted by USNOF and given system wide dissemination.

on course—

- Used to indicate that an aircraft is established on the route centerline.
- Used by ATC to advise a pilot making a radar approach that his aircraft is lined up on the final approach course.

on course indication—An indication on an instrument, which provides the pilot a visual means of determining that the aircraft is located on the centerline of a given navigational track, or an indication on a radar scope that an aircraft is on a given track.

option approach—An approach requested and conducted by a pilot which will result in either a touch-and-go, missed approach, stop-and-go, or full stop landing. (Refer to AIM.)

outer marker/OM—A marker beacon at or near the glide slope intercept altitude of an ILS approach. It is keyed to transmit two dashes per second on a 400 Hz tone, which is received aurally and visually by compatible airborne equipment. The OM is normally located four to seven miles from the runway threshold on the extended centerline of the runway. (Refer to AIM.)

parallel runways—Two or more runways at the same airport whose centerlines are parallel. In addition to runway number, parallel runways are designated as L (left) and R (right) or, if three parallel runways exist, L (left), C (center), and R (right).

pilot weather report/PIREP—A report of meteorological phenomena encountered by aircraft in flight. (Refer to AIM.)

practice instrument approach—An instrument approach procedure conducted by a VFR or an IFR aircraft for the purpose of pilot training or proficiency demonstrations.

precision approach procedure/precision approach—A standard instrument approach procedure in which an electronic glideslope/glidepath is provided; e.g., ILS/MLS and PAR.

precision approach radar/PAR—Radar equipment in some ATC facilities operated by the FAA and/or the military services at joint-use civil/military locations and separate military installations to detect and display azimuth, elevation, and range of aircraft on the final approach course to a runway. This equipment may be used to monitor certain non-radar approaches, but is primarily used to conduct a precision instrument approach (PAR) wherein the controller issues guidance instructions to the pilot based on the aircraft's position in relation to the final approach course (azimuth), the glidepath (elevation), and the distance (range) from the touchdown point on the runway as displayed on the radar scope. (Refer to AIM.)

preferential routes—Preferential routes (PDRs, PARs, and PDARs) are adapted in ARTCC computers to accomplish inter/intrafacility controller coordination and to assure that flight data is posted at the proper control positions. Locations having a need for these specific inbound and outbound routes normally publish such routes in local facility bulletins, and their use by pilots minimizes flight plan route amendments. When the workload or traffic situation permits, controllers normally provide radar vectors or assign requested routes to minimize circuitous routing. Preferential routes are usually confined to one ARTCC's area and are referred to by the following names or acronyms:

- Preferential Departure Route/PDR—A specific departure route from an airport or terminal area to an en route point where there is no further need for flow control. It may be included in a Standard Instrument Departure (SID) or a Preferred IFR Route.
- Preferential Arrival Route/PAR—A specific arrival route from an appropriate en route point to an airport or terminal area. It may be included in a standard terminal arrival (STAR) or a preferred IFR route. The abbreviation ''PAR''is used primarily within the ARTCC and should not be confused with the abbreviation for Precision Approach Radar.
- Preferential Departure and Arrival Route/PDAR—A route between two terminals which are within or immediately adjacent to one ARTCC's area. PDAR's are not synonomous with preferred IFR routes but may be listed as such as they do accomplish essentially the same purpose.

preferred IFR routes—Routes established between busier airports to increase system efficiency and capacity. They normally extend through one or more ARTCC areas and are designed to achieve balanced traffic flows among high density terminals. IFR clearances are issued on the basis of these routes except when severe weather avoidance procedures or other factors dictate otherwise. Preferred IFR routes are listed in the Airport/Facility Directory. If a flight is planned to or from an area having such routes but the departure or arrival point is not listed in the *Airport/Facility Directory*, pilots may use that part of a Preferred IFR route which is appropriate for the departure or arrival point that is listed. Preferred IFR routes are correlated with SIDs and STARs and may be defined by airways, jet routes, direct routes between NAVAIDs, Waypoints, NAVAID radials/DME, or any combination thereof.

procedure turn inbound—That point of a procedure turn maneuver where course reversal has been completed and an aircraft is established inbound on the intermediate approach segment or final approach course. A report of procedure turn inbound is normally used by ATC as a position report for separation purposes.

procedure turn/PT—The maneuver prescribed when it is necessary to reverse direction to establish an aircraft on the intermediate approach segment or final approach course. The outbound course, direction of turn, distance within which the turn must be completed, and minimum altitude are specified in the procedure. However, unless otherwise restricted, the point at which the turn may be commenced and the type and rate of turn are left to the discretion of the pilot.

radar advisory—The provision of advice and information based on radar observations.

radar approach—An instrument approach procedure which utilizes precision approach radar (PAR) or airport surveillance radar (ASR). (Refer to AIM.)

radar approach control facility—A terminal ATC facility that uses radar and nonradar capabilities to provide approach control services to aircraft arriving, departing, or transiting airspace controlled by the facility (see approach control service). Provides radar ATC services to aircraft operating in the vicinity or one of more civil and/or military airports in a terminal area. The facility may provide services of a ground controlled approach (GCA); i.e., ASR and PAR approaches. A radar approach control facility may be operated by FAA, USAF, U.S. Army, USN, USMC, or jointly by FAA and a military service.

radar contact—
- Used by ATC to inform an aircraft that it is identified on the radar display and radar flight following will be provided until radar identification is terminated. Radar service may also be provided within the limits of necessity and capability. When a pilot is informed of "radar contact," he automatically discontinues reporting over compulsory reporting points. (Refer to AIM.)
- The term used to inform the controller that the aircraft is identified and approval is granted for the aircraft to enter the receiving controllers airspace.

radar contact lost—Used by ATC to inform a pilot that radar identification of his aircraft has been lost. The loss may be attributed to several things including the aircraft's merging with weather or ground clutter, the aircraft's flying below radar line of sight, the aircraft's entering an area of poor radar return, or a failure of the aircraft transponder or the ground radar equipment.

radar environment—An area in which radar service may be provided

radar flight following—The observation of the progress of radar identified aircraft, whose primary navigation is being provided by the pilot, wherein the controller retains and correlates the aircraft identity with the appropriate target or target symbol displayed on the radar scope. (Refer to AIM.)

radar identification—The process of ascertaining that an observed radar target is the radar return from a particular aircraft.

radar service—A term which encompasses one or more of the following services based on the use of radar which can be provided by a controller to a pilot of a radar identified aircraft.

- Radar monitoring—The radar flight-following of aircraft, whose primary navigation is being performed by the pilot, to observe and note deviations from its authorized flight path, airway, or route. When being applied specifically to radar monitoring of instrument approaches; i.e., with precision approach radar (PAR) or radar monitoring of simultaneous ILS/MLS approaches, it includes advice and instructions whenever an aircraft nears or exceeds the prescribed PAR safety limit or simultaneous ILS/MLS no transgression zone.
- Radar navigational guidance—Vectoring aircraft to provide course guidance.
- Radar separation—Radar spacing of aircraft in accordance with established minima.

radar service terminated—Used by ATC to inform a pilot that he will no longer be provided any of the services that could be received while in radar contact. Radar service is automatically terminated, and the pilot is not advised in the following case:

- An aircraft cancels its IFR flight plan, except within a TCA, TRSA, ARSA, or where Stage II service is provided.
- An aircraft conducting an instrument, visual, or contact approach has landed or has been instructed to change to advisory frequency.
- An arriving VFR aircraft, receiving radar service to a tower-controlled airport within a TCA, TRSA, ARSA, or where Stage II service is provided, has landed; or to all other airports, is instructed to change to tower or advisory frequency.
- An aircraft completes a radar approach.

radar weather echo intensity levels—Existing radar systems cannot detect turbulence. However, there is a direct correlation between the degree of turbulence and other weather features associated with thunderstorms and the radar weather echo intensity. The National Weather Service has categorized six levels of radar weather echo intensity. The levels are sometimes expressed during communications as "VIP LEVEL" 1 through 6 (derived from the component of the weather radar that produces the information—Video Integrator and Processor). The following list gives the weather features likely to be associated with these levels during thunderstorm weather situations:

- LEVEL 1 (weak) and LEVEL 2 (moderate). Light to moderate turbulence is possible with lightning.
- LEVEL 3 (strong). Severe turbulence possible, lightning.
- LEVEL 4 (very strong). Severe turbulence likely lightning.
- LEVEL 5 (intense). Severe turbulence, lightning, organized wind gusts, hail likely.
- LEVEL 6 (extreme). Severe turbulence, large hail, lightning, extensive wind gusts, and turbulence.

radial—A magnetic bearing extending from a VOR/VORTAC/TACAN navigation facility.

radio altimeter/radar altimeter—Aircraft equipment which makes use of the reflection of radio waves from the ground to determine the height of the aircraft above the surface.

radio magnetic indicator/RMI—An aircraft navigational instrument coupled with a gyro compass or similar compass that indicates the direction of a selected NAVAID and indicates bearing with respect to the heading of the aircraft.

remote communications air/ground facility/RCAG—An unmanned VHF/UHF transmitter/ receiver facility which is used to expand ARTCC air/ground communications coverage and to facilitate direct contact between pilots and controllers. RCAG facilities are sometimes not equipped with emergency frequencies 121.5 MHz and 243.0 MHz. (Refer to AIM.)

remote communications outlet/RCO and remote transmitter/receiver/RTR—An unmanned communications facility remotely controlled by air traffic personnel. RCOs serve FSSs. RTRs serve terminal ATC facilities. An RCO or RTR may be UHF or VHF and will extend the communication range of the air traffic facility. There are several classes of RCOs and RTRs. The class is determined by the number of transmitters or receivers. Classes A through G are used primarily for air/ground purposes. RCO and RTR class O facilities are nonprotected outlets subject to undetected and prolonged outages. RCOs and RTRs were established for the express purpose of providing ground-to-ground communications between air traffic control specialists and pilots located at a satellite airport for delivering en route clearances, issuing departure authorizations, and acknowledging instrument flight rules cancellations or departure/landing times. As a secondary function, they may be used for advisory purposes whenever the aircraft is below the coverage of the primary air/ground frequency.

request full route clearance/FRC—Used by pilots to request that the entire route of flight be read verbatim in an ATC clearance. Such request should be made to preclude receiving an ATC clearance based on the original filed flight plan when a filed IFR flight plan has been revised by the pilot, company, or operations prior to departure.

resume own navigation—Used by ATC to advise a pilot to resume his own navigational responsibility. It is issued after completion of a radar vector or when radar contact is lost while the aircraft is being radar vectored.

roger—I have received all of your last transmission. It should not be used to answer a question requiring a yes or a no answer.

route—A defined path, consisting of one or more courses in a horizontal plane, which aircraft traverse over the surface of the earth. (See airway, jet route, published route, unpublished route.)

route segment—As used in air traffic control, a part of a route that can be defined by two navigational fixes, two NAVAIDs, or a fix and a NAVAID.

say again—Used to request a repeat of the last transmission. Usually specifies transmission or portion thereof not understood or received; e.g., "Say again all after ABRAM VOR."

say altitude—Used by ATC to ascertain an aircraft's specific altitude/flight level. When the aircraft is climbing or descending, the pilot should state the indicated altitude rounded to the nearest 100 feet.

say heading—Used by ATC to request an aircraft heading. The pilot should state the actual heading of the aircraft.

segments of an instrument approach procedure—An instrument approach procedure may have as many as four separate segments depending on how the approach procedure is structured.

1. Initial approach—The segment between the initial approach fix and the intermediate fix or the point where the aircraft is established on the intermediate course or final approach course.

2. Intermediate approach—The segment between the intermediate fix or point and the final approach fix.

3. Final approach—The segment between the final approach fix or point and the runway, airport, or missed approach point.

4. Missed approach—The segment between the missed approach point or the point of arrival at decision height and the missed approach fix at the prescribed altitude. (Refer to FAR, Part 97.)

separation—In air traffic control, the spacing of aircraft to achieve their safe and orderly movement in flight and while landing and taking off.

severe weather avoidance plan/SWAP—An approved plan to minimize the affect of severe weather on traffic flows in impacted terminal and/or ARTCC areas. SWAP is normally implemented to provide the least disruption to the ATC system when flight through portions of airspace is difficult or impossible due to severe weather.

severe weather forecast alerts/AWW—Preliminary messages issued in order to alert users that a Severe Weather Watch Bulletin (WW) is being issued. These messages define areas of possible severe thunderstorms or tornado activity. The messages are unscheduled and issued as required by the National Severe Storm Forecast Center at Kansas City, Missouri.

SIGMET/WS/significant meteorological information—A weather advisory issued concerning weather significant to the safety of all aircraft. SIGMET advisories cover severe and extreme turbulence, severe icing, and widespread dust or sandstorms that reduce visibility to less than 3 miles. (Refer to AIM.)

simplified directional facility/SDF—A NAVAID used for nonprecision instrument approaches. The final approach course is similar to that of an ILS localizer except that the SDF course may be offset from the runway, generally not more than 3 degrees, and the course may be wider than the localizer, resulting in a lower degree of accuracy. (Refer to AIM.)

simultaneous ILS/MLS approaches—An approach system permitting simultaneous ILS/MLS approaches to airports having parallel runways separated by at least 4,300 feet between centerlines. Integral parts of a total system are ILS/MLS, radar, communications, ATC procedures, and appropriate airborne equipment. (Refer to AIM.)

special VFR conditions—Weather conditions in a control zone which are less than basic VFR and in which some aircraft are permitted flight under Visual Flight Rules. (Refer to FAR, Part 91.)

special VFR operations—Aircraft operating in accordance with clearances within control zones in weather conditions less than the basic VFR weather minima. Such operations must be requested by the pilot and approved by ATC.

standard instrument departure/SID—A preplanned instrument flight rule (IFR) air traffic control departure procedure printed for pilot use in graphic and/or textual form. SIDs provide transition from the terminal to the appropriate en route structure. (Refer to AIM.)

standard rate turn—A turn of three degrees per second.

standard terminal arrival/STAR—A preplanned instrument flight rule (IFR) air traffic control arrival procedure published for pilot use in graphic and/or textual form. STARs provide transition from the en route structure to an outer fix or an instrument approach fix/arrival waypoint in the terminal area.

stand by—Means the controller or pilot must pause for a few seconds, usually to attend to other duties of a higher priority. Also means to wait as in "stand by for clearance." If a delay is lengthy, the caller should reestablish contact.

stepdown fix—A fix permitting additional descent within a segment of an instrument approach procedure by identifying a point at which a controlling obstacle has been safely overflown.

stop altitude squawk—Used by ATC to inform an aircraft to turn-off the automatic altitude reporting feature of its transponder. It is issued when the verbally reported altitude varies 300 feet or more from the automatic altitude report.

substitute route—A route assigned to pilots when any part of an airway or route is unusable because of NAVAID status.

sunset and sunrise—The mean solar times of sunset and sunrise as published in the Nautical Almanac, converted to local standard time for the locality concerned. Within Alaska, the end of evening civil twilight and the beginning of morning civil twilight, as defined for each locality.

surveillance approach—An instrument approach wherein the air traffic controller issued instructions, for pilot compliance, based on aircraft position in relation to the final approach course (azimuth), and the distance (range) from the end of the runway as displayed on the controller's radar scope. The controller will provide recommended altitudes on final approach if requested by the pilot. (Refer to AIM.)

target—The indication shown on a radar display resulting from a primary radar return or a radar beacon reply.

terminal radar program—A national program instituted to extend the terminal radar services provided IFR aircraft to VFR aircraft. Pilot participation in the program is urged but is not mandatory. The program is divided into two parts and referred to as

Stage II and Stage III. The Stage service provided at a particular location is contained in the *Airport/Facility Directory.*

1. Stage I originally comprised two basic radar services (traffic advisories and limited vectoring to VFR aircraft). These services are provided by all commissioned terminal radar facilities, but the term "Stage I" has been deleted from use.
2. State II/Radar Advisory and Sequencing for VFR Aircraft—Provides, in addition to the basic radar services, vectoring and sequencing on a full-time basis to arriving VFR aircraft. The purpose is to adjust the flow of arriving IFR and VFR aircraft into the traffic pattern in a safe and orderly manner and to provide traffic advisories to departing VFR aircraft.
3. Stage III/Radar Sequencing and Separation Service for VFR Aircraft—Provides, in addition to the basic radar services and Stage II, separation between all participating VFR aircraft. The purpose is to provide separation between all participating VFR aircraft and all IFR aircraft operating within the airspace defined as a terminal radar service area (TRSA) or terminal control area (TCA). (Refer to AIM, *Airport/Facility Directory.*)

terminal radar service area/TRSA—Airspace surrounding designated airports wherein ATC provides radar vectoring, sequencing, and separation on a full-time basis for all IFR and participating VFR aircraft. Service provided in a TRSA is called Stage III Service. The AIM contains an explanation of TRSA. TRSAs are depicted on VFR aeronautical charts. Pilot participation is urged but is not mandatory. (Refer to AIM, Airport/Facility Directory.)

touchdown zone—The first 3,000 feet of the runway beginning at the threshold. The area is used for determination of touchdown zone elevation in the development of straight-in landing minimums for instrument approaches.

tower en route control service/tower to tower—The control of IFR en route traffic within delegated airspace between two or more adjacent approach control facilities. This service is designed to expedite traffic and reduce control and pilot communication requirements.

track—The actual flight path of an aircraft over the surface of the earth.

traffic advisories—Advisories to alert pilots to other known or observed air traffic which may be in such proximity to the position of intended route of flight of their aircraft to warrant their attention. Such advisories may be based on:
• Visual observation.
• Observation of radar identified and nonidentified aircraft targets on an ATC radar display, or
• Verbal reports from pilots or other facilities. The word "traffic" followed by additional information, if known, is used to provide such advisories; e.g., "Traffic, 2 o'clock, one zero miles, southbound, eight thousand." Traffic advisory service will be provided to the extent possible depending on higher priority duties of the controller or other limitations; e.g., radar limitations, volume of traffic, frequency

congestion, or controller workload. Radar/nonradar traffic advisories do not relieve the pilot of his responsibility to see and avoid other aircraft. Pilots are cautioned that there are many times when the controller is not able to give traffic advisories concerning all traffic in the aircraft's proximity; in other words, when a pilot requests or is receiving traffic advisories, he should not assume that all traffic will be issued. (Refer to AIM, Radar Traffic Information Service.)

transfer of control—That action whereby the responsibility for the separation of an aircraft is transferred from one controller to another.

transmissometer—An apparatus used to determine visibility by measuring the transmission of light through the atmosphere. It is the measurement source for determining runway visual range (RVR) and runway visibility value (RVV). (See visibility.)

transmitting in the blind/blind transmission—A transmission from one station to other stations in circumstances where two-way communication cannot be established, but where it is believed that the called stations may be able to receive the transmission.

transponder—The airborne radar beacon receiver/transmitter portion of the air traffic control radar beacon system (ATCRBS) which automatically receives radio signals from interrogators on the ground, and selectively replies with a specific reply pulse or pulse group only to those interrogations being received on the mode to which it is set to respond. (Refer to AIM.)

T-VOR/terminal-very high frequency omnidirectional range station—A very high frequency terminal omnirange station located on or near an airport and used as an approach aid.

unable—Indicates inability to comply with a specific instruction, request, or clearance.

under the hood—Indicates that the pilot is using a hood to restrict visibility outside the cockpit while simulating instrument flight. An appropriately rated pilot is required in the other control seat while this operation is being conducted. (Refer to FAR, Part 91.)

UNICOM—A nongovernment communication facility which may provide airport information at certain airports. Locations and frequencies of UNICOMs are shown on aeronautical charts and publications. (Refer to AIM, *Airport/Facility Directory.*)

vector—A heading issued to an aircraft to provide navigational guidance by radar.

verify—Request confirmation of information; e.g., "verify assigned altitude."

vertical separation—Separation established by assignment of different altitudes or flight levels.

very high frequency/VHF—The frequency band between 30 and 300 MHz. Portions of this band, 108 to 118 MHz, are used for certain navaids; 118 to 136 MHz are used for civil air/ground voice communications. Other frequencies in this band are used for purposes not related to air traffic control.

VFR-on-top—ATC authorization for an IFR aircraft to operate in VFR conditions at any appropriate VFR altitude (as specified in FAR and as restricted by ATC). A pilot receiving this authorization must comply with the VFR visibility, distance from cloud criteria, and the minimum IFR altitudes specified in FAR Part 91. The use of this term does not relieve controllers of their responsibility to separate aircraft in TCAs/TRSAs as required by FAA Handbook 7110.65.

visibility—The ability, as determined by atmospheric conditions and expressed in units of distance, to see and identify prominent unlighted objects by day and prominent lighted objects by night. Visibility is reported as statute miles, hundreds of feet or meters. (Refer to FAR Part 91, AIM.)

- Flight visibility—The average forward horizontal distance, from the cockpit of an aircraft in flight, at which prominent unlighted objects may be seen and identified by day and prominent lighted objects may be seen and identified by night.
- Ground visibility—Prevailing horizontal visibility near the earth's surface as reported by the United States National Weather Service or an accredited observer.
- Prevailing visibility—The greatest horizontal visibility equaled or exceeded throughout at least half the horizon circle which need not necessarily be continuous.
- Runway visibility value/RVV—The visibility determined for a particular runway by a transmissometer. A meter provides a continuous indication of the visibility (reported in miles or fractions of miles) for the runway. RVV is used in lieu of prevailing visibility in determining minimums for a particular runway.
- Runway visual range/RVR—An instrumentally derived value, based on standard calibrations, that represents the horizontal distance a pilot will see down the runway from the approach end. It is based on the sighting of either high intensity runway lights or on the visual contrast of other targets whichever yields the greater visual range. RVR, in contrast to prevailing or runway visibility, is based on what a pilot in a moving aircraft should see looking down the runway. RVR is horizontal visual range, not slant visual range. It is based on the measurement of a transmissometer made near the touchdown point of the instrument runway and is reported in hundreds of feet. RVR is used in lieu of RVV and/or prevailing visibility in determining minimums for a particular runway. (Touchdown RVR—The RVR visibility readout values obtained from RVR equipment serving the runway touchdown zone.) (Mid-RVR—The RVR readout values obtained from RVR equipment located midfield of the runway.) (Rollout RVR—The RVR readout values obtained from RVR equipment located nearest the rollout end of the runway.)

visual approach—An approach wherein an aircraft on an IFR flight plan, operating in VFR conditions under the control of an air traffic control facility and having an air traffic control authorization, may proceed to the airport of destination in VFR conditions.

visual descent point/VDP—A defined point on the final approach course of a nonprecision straight-in approach procedure from which normal descent from the MDA to the runway touchdown point may be commenced, provided the approach threshold of that runway, or approach lights, or other markings identifiable with the approach end of that runway are clearly visible to the pilot.

visual separation—A means employed by ATC to separate aircraft in terminal areas. There are two ways to effect this separation:

- The tower controller sees the aircraft involved and issues instructions, as necessary, to ensure that the aircraft avoid each other.
- A pilot sees the other aircraft involved and upon instructions from the controller provides his own separation by maneuvering his aircraft as necessary to avoid it. This may involve following another aircraft or keeping it in sight until it is no longer a factor. (Refer to FAR, Part 91.)

VORTAC/VHF omnidirectional range/tactical air navigation—A navigation aid providing VOR azimuth, TACAN azimuth, and TACAN distance measuring equipment (DME) at one site. (Refer to AIM.)

vortices/wing tip vortices—Circular patterns of air created by the movement of an airfoil through the air when generating lift. As an airfoil moves through the atmosphere in sustained flight, an area of low pressure is created above it. The air flowing from the high pressure area to the low pressure area around and about the tips of the airfoil tends to roll up into two rapidly rotating vortices, cylindrical in shape. These vortices are the most predominant parts of aircraft wake turbulence and their rotational force is dependent upon the wing loading, gross weight, and speed of the generating aircraft. The vortices from medium to heavy aircraft can be of extremely high velocity and hazardous to smaller aircraft. (See aircraft classes wake turbulence.) (Refer to AIM.)

VOT/VOR test signal—A ground facility which emits a test signal to check VOR receiver accuracy. Some VOTs are available to the user while airborne, and others are limited to ground use only. (Refer to FAR, Part 91, AIM, *Airport/Facility Directory.*)

wake turbulence—Phenomena resulting from the passage of an aircraft through the atmosphere. The term includes vortices, thrust stream turbulence, jet blast, jet wash, propeller wash, and rotor wash both on the ground and in the air. (Refer to AIM.)

waypoint—A predetermined geographical position used for route/instrument approach definition, or progress reporting purposes, that is defined relative to a VORTAC station or in terms of latitude/longitude coordinates.

wilco—I have received your message, understand it, and will comply with it.

wind shear—A change in wind speed and/or wind direction in a short distance resulting in a tearing or shearing effect. It can exist in a horizontal or vertical direction and occasionally in both.

Index